Acclaim for Katie Spalding's

EDISON'S GHOSTS

THE UNTOLD WEIRDNESS OF HISTORY'S GREATEST GENIUSES

"Katie Spalding is a kind of Bill Nye on steroids, making arcane science fascinating and fun…Funny, and with an absurd amount of obscure knowledge, *Edison's Ghosts* is a must-read on how everyone is much, much stupider than they make out."
—James Felton, author of *Assholes: The Dead People You Should Be Mad At*

"With wit and charm, each of Katie Spalding's stories nudges, pushes, and eventually shoves some of our most illustrious celebrity thinkers right off their pedestals. Whether it was learning how Pythagoras died from an ill-timed fascination with beans, the career-derailing procrastination of Leonardo Da Vinci, the truly impressive-in-its-magnitude gullibility of Arthur Conan Doyle, or the failed attempt of the titular Edison to create a phone for calling ghosts, this warts-and-all review of the human—the very silly human—side of history's most famous 'geniuses' will fuel your dinner party conversations for years."
—David McRaney, author of *You Are Not So Smart*

"*Edison's Ghosts* is a masterful combination of historical research and comedic storytelling, infused with erudition and judiciously dropped F-bombs. I laughed out loud on nearly every page. It is truly inspiring to read about the stupidity of geniuses. Thank you, Katie, for knocking these wunderkinds down a few pegs and making the rest of us feel smarter in the process."
—Justin Gregg, author of *If Nietzsche Were a Narwhal*

"A lighthearted and amusing account of some of history's most influential people. Even the brightest minds can produce some truly dim moments, and this book doesn't hold back."
—Nick Caruso, *New York Times* bestselling author of *Does it Fart?*

"Spalding delivers consistently lively, witty excursions into the sometimes-weird lives and beliefs of the famous, from Pythagoras to Yukio Mishima... An entertaining and informative collection."
—*Kirkus Reviews*

"An extremely entertaining book... The Leonardo chapter left me helpless with mirth and the Karl Marx one brilliantly tells my favorite 'Karl Marx gets pissed' anecdote better than any version I've ever seen. Smart and hilarious." —Otto English, author of *Fake History*

"Such a great idea I wish I'd come up with it first. One of those books that makes you laugh so much you forget you're learning stuff."
—Jonn Elledge, author of *The Compendium of (Not Quite) Everything*

"Until now I thought 'enjoyable science book' was an oxymoron. Spalding proved me wrong. I learned a lot and had fun doing it. Turns out a spoonful of snark helps the factoids go down—in a most delightful way." —Curt Schleier, *Minneapolis Star Tribune*

"There's something hopeful about a book setting out to tell us that our fabled geniuses were weirdos and goofballs. It supposes that greatness is still real enough to need demystifying. Spalding teases her subjects because, most of the time, she wants to express what was admirable about them." —*Air Mail*

EDISON'S GHOSTS

THE UNTOLD WEIRDNESS OF HISTORY'S GREATEST GENIUSES

KATIE SPALDING

BACK BAY BOOKS
Little, Brown and Company
New York Boston London

Copyright © 2023 by Katie Spalding

Hachette Book Group supports the right to free expression and the value of copyright. The purpose of copyright is to encourage writers and artists to produce the creative works that enrich our culture.

The scanning, uploading, and distribution of this book without permission is a theft of the author's intellectual property. If you would like permission to use material from the book (other than for review purposes), please contact permissions@hbgusa.com. Thank you for your support of the author's rights.

Back Bay Books / Little, Brown and Company
Hachette Book Group
1290 Avenue of the Americas, New York, NY 10104
littlebrown.com

Originally published in hardcover by Little, Brown and Company, May 2023

First Back Bay paperback edition, August 2024

Back Bay Books is an imprint of Little, Brown and Company, a division of Hachette Book Group, Inc. The Back Bay Books name and logo are trademarks of Hachette Book Group, Inc.

The publisher is not responsible for websites (or their content) that are not owned by the publisher.

The Hachette Speakers Bureau provides a wide range of authors for speaking events. To find out more, go to hachettespeakersbureau.com or email HachetteSpeakers@hbgusa.com.

Little, Brown and Company books may be purchased in bulk for business, educational, or promotional use. For information, please contact your local bookseller or the Hachette Book Group Special Markets Department at special.markets@hbgusa.com.

Chapter opener line drawings courtesy of Shutterstock

ISBN 978-0-316-52952-5 (hc) / 978-0-316-52954-9 (pb)

LCCN is available from the Library of Congress

Printing 2, 2025

LSC-C

Printed in the United States of America

To my grandad and my Roo

Contents

Introduction   1

1. The Mathematical Cult Leader Pythagoras, and His Incredibly Stupid Death   9

2. Confucius Was an Ugly Nerd with Low Self-Esteem   18

3. Never, Ever Hire Leonardo da Vinci   26

4. Galileo Utterly Fails to Read the Room   36

5. The Entirely Unbelievable Life of Tycho Brahe   49

6. When René Descartes Got Baked   62

7. Isaac Newton and the Philosopher's Stone   72

8. Mozart Uses His Superstar Status to Tell Us All to Kiss His Arse ... Over and Over Again   84

9. Benjamin Franklin Uses World-Changing Technology to Prank Friends, Self — 93

10. Émilie du Châtelet Cares Not for Your Social Mores, and She Will Fight You in Her Underwear to Prove It — 100

11. Johann Christian Reil Invents Psychiatry and Things Get Really Weird Really Quickly — 110

12. Napoleon Bonaparte's Fluffiest Foe — 120

13. Lord Byron, the Patron Saint of Fuckboys — 125

14. Ada Lovelace's (Husband's) Family Jewels — 137

15. Galois Hunting — 144

16. John Couch Adams Ignores His Mail, Loses Neptune — 154

17. You Really Wouldn't Want to Hang Out with Karl Marx — 165

18. Charles Darwin: Glutton; Worm Dad; Murderer? — 174

19. James Glaisher, the Victorian Weatherman Who Nearly Became an Astronaut — 183

20. Sigmund Freud Used Cocaine So Much He Thought Numbers Wanted to Kill Him — 195

21. Arthur Conan Doyle Gets Pranked So Hard He Claims Fairies Exist — 206

22. Thomas Edison's Lesser-Known Invention: Dial-a-Ghost — 219

23.	Real-Life Supervillain Nicola Tesla Takes the Term 'Pigeon Fancying' a Bit Too Literally	229
24.	Marie Curie Defies All the Odds to Accidentally Poison Both Herself and Thousands of Strangers	235
25.	Albert Einstein: Public Nuisance, Love Rat	242
26.	Kurt Gödel, the Disney Princess Who Broke Time	254
27.	Maya Angelou, in: Stop! Or My Mom Will Shoot	263
28.	Ernest Hemingway May Have Been the Worst Double Agent Ever	274
29.	Yukio Mishima and the Shortest, Gayest Fascist Coup in History	284
30.	NASA Forgets about Women, Toilets and the Metric System	293
Epilogue		305
References		307
Acknowledgements		333
Index		335

Introduction

As Albert Einstein almost certainly never said, everyone is a genius – but if you judge a fish by its ability to climb a tree, it will live its whole life believing that it is stupid. The message behind it is clear: everyone has different strengths, and you shouldn't feel bad that you have no idea how c stands for 'speed of light' or why multiplying it by m will give you E, because there's probably a bunch of stuff you can do that Einstein couldn't.

It's kind of ironic that whoever first came up with this saying chose Einstein as its supposed progenitor, because in fish terms, the man who gave the world the theories of special and general relativity, rewriting the very laws of physics of the universe, was dumb as a box of rocks.

Let me explain: imagine you're a fish, living in the Atlantic Ocean off the coast of Long Island in the late 1930s. There you are, swimming along with all your fishy pals, minding your fishy business – when, suddenly, in bursts some huge hairy beast out of the sky, splashing around, making a huge scene and freaking everybody out with its cries of 'Help! Help! I can't swim, I'm going to drown!' Which you don't even understand, because, you know, you're a fish.

And now imagine that's the sixth time that's happened. This week.

At some point, you're going to stop swimming away in fear from this wild-haired intruder and start trying to put it out of its obvious misery. I guess what I'm saying is: it's amazing that Einstein was never murdered by a bunch of pissed-off cod, because despite being probably the guy most people in the West would associate with the word 'genius', he never learned to swim.

Now, I'm not here to tell you that's some kind of fatal personality flaw – plenty of perfectly lovely people can't swim. But those people don't tend to spend their summers pursuing a hobby where an inability to swim can literally result in your death. Not so with Einstein, who was an enthusiastic and objectively fucking terrible ocean sailor, and whose reputation in the hamlet of Cutchogue, New York, is to this day less 'epoch-defining physicist' and more 'public nuisance who wouldn't stop forcing us all to jump in the sea and save him'.

See, the flipside of 'everyone is brilliant in their own way' is the equally true 'everyone is an idiot', and that seems to be particularly true when we talk about the people traditionally held up as 'geniuses'. Maybe it's just the apparent contrast between what we expect from these figures and what we get. For example, it probably wouldn't be totally surprising to learn that, say, Chris, the kid you went to school with who won the 'gave the most teachers a mental breakdown' award at prom, is now being sued for failing to deliver on some construction work, but that's just not the sort of behaviour you'd expect from Leonardo da Vinci. He's meant to be ... you know ... *better* than us.

Then again. They say we only use 10 per cent of our brains at any one time, which just goes to show how much 'they' know about neuroscience – but really, should we expect people who are busy literally changing the world to also be able to concentrate on more prosaic things, like contract law and swimming lessons?

The germ of the idea for this project came to me a few years ago, in the aftermath of a total solar eclipse. I was in Durham at the time; I was about halfway through my maths PhD, and the university was holding its annual summer Mathematical Symposium – a wonderful opportunity for students and professors to meet, exchange ideas, collaborate and, most importantly, get ludicrously drunk. You might not believe it, but there are very few people in my experience who can party harder than a bunch of professional mathematicians.

The eclipse was a pretty amazing sight – I remember everyone stepping out of whichever seminar was going on at the time to watch as the sun disappeared behind the moon for a few precious seconds. Evidently, awe-inspiring cosmological phenomena are more interesting than the finer points of Teichmüller theory for some people. Takes all sorts, I suppose.

But it wasn't the eclipse *itself* that people were talking and reading and memeing about in the following days. What had caught the imagination of the world at large wasn't the grand majesty of nature, but the folly of man – specifically, a man named President Donald J. Trump.

Despite this man having access to resources like secret service agents dedicated to keeping him safe, and top government scientists and advisers, and presumably at least a primary-school-level education, the publicity shots that came out of the White House that year to commemorate the eclipse showed Trump very clearly squinting, bare-eyed, right up at the sun.

Immediately, people started asking the obvious question: was this man really so stupid as to stare directly at the sun? And, I mean, yes, he was, but it led me to think of someone else. Because you know who else was that stupid? Isaac Newton. Yes, *that* Newton: the guy with the apple who we all, unquestioningly, think of as one of history's greatest geniuses.

In fact, Newton arguably outdid Trump by several levels of stupidity. He didn't just look at the sun without eye protection – he put that famously big and inventive brain to work setting up the precise scenario that would cause the maximum damage to his eyes possible. The result? Three days of blindness, and what is possibly the most revolutionary investigation into light and optics ever written.

Now, as far as I'm aware, Donald Trump has not yet given the world an epoch-defining scientific treatise. Maybe one day he will, and staring directly at the sun will come to be seen as just one of those things geniuses do, like playing the violin or getting brained by apples. But for me, the fact that my friends' disbelieving cries of 'who the fuck doesn't know not to stare directly at the sun?!' could be smugly replied to with 'Isaac Newton, actually' was just incredibly funny, and since there's only so many times you can tell your friends how ignorant they are before they start actively avoiding you – and Trump didn't seem likely to go ocean sailing any time soon – I decided to write it all down.

Normally, biographies of the great and the good tend to hide these weird little ventures of their subjects in a footnote or some throwaway sentence – something your eyes will skip over while you digest the more general point of how amazing they were. Surely, if only for symmetry's sake, there should be at least one book out there that reverses that rule – that tells you how dumb and relatable they were, and *maybe* imparts some incidental lesson about their achievements.

In general, I think, there are two kinds of these weirdo geniuses. The first group are the Mozarts and Confuciuses of history: people who, had they not been objectively incredible at what they did, would probably have gone down in history as 'that pervert from Getreidegasse' or 'some ugly janitor who kept asking if you wanted to read his history books'. They're the lucky bastards, basically – the ones who got away with foibles us mere mortals would be forced to save for our anonymous tumblr accounts.

The other kind are perhaps more interesting. These are people like Marie Curie, or the aforementioned Leonardo da Vinci – the ones whose shortcomings were pretty tightly interwoven with their accomplishments. Take away Leonardo's flighty imagination and you'd get – what? A guy who was much better at delivering on contracts, sure, but probably not the *Mona Lisa*. And while it's easy, as a modern reader, to scoff at Curie's habit of carrying radioactive elements around in her pockets and bare hands, you have to admit: if she wasn't the kind of woman willing to handle warm glowing lumps of metal while watching her skin inexplicably burn off in real-time because of it, then she wouldn't have been the kind of woman who could discover radium.

So who counts? After all, we've all done stupid things now and then – I myself once accidentally swallowed a bunch of that stuff you get inside a glow stick, and, inexplicably, nobody gave *me* a Nobel Prize for it. Clearly, just being an idiot isn't enough. But as I increasingly found while compiling stories of humanity's brightest and best, neither is *just* being a genius.

I want you to try something for me: type 'geniuses' into a search engine and see what comes up. I'm going to go out on a limb here and guess you got, among all the definitions and photos of Einstein, a lot of links to biographies of people like Newton, or Stephen Hawking, or Terry Tao, or even Jackie Chan, depending on the tastes of whoever compiled the list.

If you haven't noticed a problem there, then try searching for 'female geniuses' instead. Sure, you'll get some lists and biographies, but a lot of them will be called things like 'Why Haven't You Heard of These Female Geniuses?' or 'In Recognition of International Women's Day, Here Are Ten Female Geniuses You've Never Heard Of', or most egregiously, 'Why Aren't There More Women Geniuses?'

You'll likely get the same sort of thing if you search for 'Black geniuses' or 'LGBTQ geniuses', or any number of qualifiers. It's not like these people don't exist; it's not even that people aren't aware of them. But for whatever reason – and let's face it, we all know the reasons – they don't seem to count as our *default* geniuses. We in the West are mostly still recovering from centuries of essentially banning most people from stuff that you tend to need if you're going to make history – lots of free time, for instance; access to education; making it past the age of nine without dying from coal miner's lung; that kind of thing – and while absolutely some women, and minorities, and working-class people, did manage to beat the odds and change the world in their own ways, they tend to be remembered today as, well, precisely that: the ones who beat the odds.

And that's fine and wonderful and all very inspiring, but it sucks for the purposes of this book, because when we're only focusing on how amazing and wonderful somebody was for outsmarting the patriarchy or whatever, we tend to miss the bizarre little habits that made them imperfect and unique. One of my favourite figures from American history, for example, is a woman named Mary Ellen Pleasant – she was a philanthropist, a businesswoman, an anti-segregation and anti-slavery activist, and probably the first ever Black self-made millionaire, all of which she did at a point in time when walking across the wrong state line could get her legally kidnapped and forced into slavery for the rest of her life. I would have loved to include her in my list of geniuses, because, frankly, she was eminently qualified for the title, but I couldn't, because nobody outside of contemporary racists seemed to have a bad word to say about her. It's the same with, say, Srinivasa Ramanujan – an almost entirely self-taught mathematician, and someone who is truly legendary for the discoveries and observations he made in his sadly far-too-short lifetime, but as far as I can tell, not in any way an idiot. I mean, for Pete's sake, the man invented a new type of number *while he was in hospital*. He wasn't even working at the time! It was halfway through a conversation with a visiting friend!

My point is, I haven't compiled an exhaustive list of all the geniuses in history. To get into the book, the subjects needed a very particular set of attributes: smart enough to be noteworthy, but not so smart as to never do anything really fucking dumb; inspiring enough for it to be unexpected when you learn they couldn't understand some pretty basic life skills, but not so inspiring that society hasn't yet let them be flawed.

You might think, when you read my judgements on certain characters, that I've been unduly harsh – 'eviscerated', I think, was the word my mum used when I showed her the first draft. This is ... probably true, but in fairness, they did some really stupid shit, and it's long past due that somebody called Lord Byron an arsehole, in my opinion. We can only hope that, with time, the right to fuck up is extended to everybody – because if reading through the stories of the thirty or so figures who did make it on to the list teaches you anything (and let me be clear: that's not a guarantee), it should be that we're all equally guilty of doing so.

Basically, I think, the difference between you and Einstein, or between me and Mozart, is mostly one of luck. They were lucky that their brilliance outweighed their sins – and you and I are actually lucky to not be so clever as to have our flaws recorded for future snarky writers to compile in a book about how entirely we showed our own arses that one time.

I.

The Mathematical Cult Leader Pythagoras, and His Incredibly Stupid Death

If there's one thing you know about Pythagoras, and let's face it, there probably is, it's this:

$$a^2 + b^2 = c^2$$

Doesn't look familiar? Maybe you know it in its more literary form: 'the square of the hypotenuse is equal to the sum of the squares of the other two sides'. It's a simple little theorem that's been around for millennia, and it's been incredibly important throughout that entire time – though it may not seem like it at first glance, since it's most often seen in its raw form in GCSE maths textbooks filled with miniature diagrams of triangles.

And that's why, if you've ever thought about who Pythagoras actually was as a person – you know, before he found lasting fame as an equation – you probably assumed he was a mathematician, and not necessarily a very good one. And that's … true, actually, for at least a couple of reasons. But it's not the *whole* truth.

If we could hop in some kind of time-and-space machine and go back to Ancient Greece in the sixth century BCE, we could actually ask Pythagoras what he considered himself to be. And he would probably say, 'I'm a philosopher.'

Actually, scratch that. He'd probably say, 'How dare you talk to me, you worthless serf. Death! Death for one thousand years!! Worshippers, take them away!'

We don't actually know all that many facts about the life of Pythagoras. Much like Jesus, or Batman, so much mythology has grown up around him over the years that it's nigh-impossible at this point to separate fact from fiction, and even fans of the guy will openly admit that pretty much any detail in a biography of him will be contradicted by a different one.

One thing we can say for sure, though, is that he must have had just *ridiculous* levels of charisma, since there's simply no way any normal person could get away with some of the bullshit he pulled. The very fact that we think of him as a mathematician is testament to this, since it's fairly doubtful he actually contributed much to mathematics at all. Even the theorem that bears his name was originally discovered a full millennium before he was born, meaning that, yes, one of the most famous pieces of mathematics is nothing more than an Ancient Greek marketing trick.

See, what Pythagoras was above all else was an *innovator*. He called himself a philosopher, but 'philosopher' was just a word he made up himself, presumably after somebody else came up with the concept of a CV and started questioning whether 'just sitting around thinking about stuff, I guess' was a good thing to list under 'previous experience'. And much more than some silly rule about triangles, we ought to be thanking Pythagoras for that other mainstay of the modern world that he created: the mad cult, shacking up in the compound down the road with all those tie-dyed vegans and the posters about how only by joining their church can you discover the mysteries of the universe.

Follow Pythagoras and you could learn such revelations as 'the Earth isn't at the centre of the universe; it actually orbits an enormous burning fire in the middle of the solar system' and 'women are as good at being academics as men are, and we should let them go to school', both of which are true and actually quite impressively ahead of their time as philosophies go.* Of course, you'd also learn things like 'stars are musical and the reason you don't know about it is because you're so used to the constant noise they make' and 'there's a secret other Earth that is constantly on the other side of the Central Fire and that's why the universe doesn't tip over', which are less true, but you can't win 'em all, I guess.

The thing is, the Pythagoreans were – and this is putting it mildly – worryingly gullible. They may have been the first people to believe in Australia, but they also thought their leader could time travel and that he could recognise his dead friends after they'd been reincarnated as dogs, which are two abilities with arguably very different levels of usefulness. Even Aristotle, a dude whose reputation for cleverness has possibly been overstated, went on the record to say Pythagoras could talk to water, be in two places at once, and had a thigh made of pure gold, like some kind of Disney's Pocahontas-Time Lord-C-3PO hybrid.

They also lived, as any good cultist worth their salt must, by some pretty bizarre and unjustifiably strict commandments. So, for example, much like modern philosophy undergraduates, new Pythagoreans had to give up all their possessions and live in a commune, and spend five years listening in total silence to strange lectures delivered by somebody hiding behind a curtain.

But less like the modern university experience was the rule that finding out the identity of this mystery lecturer would result in being

---

* To be clear, this 'Central Fire' did not refer to the sun, which the Pythagoreans taught was also orbiting, further out from Earth. This does make the theory *somewhat* less accurate, but you have to bear in mind that simply suggesting that the Earth wasn't at the centre of the universe was a pretty big deal at this point, so they get marks for originality, if nothing else.

declared dead by the group. And I say 'declared dead', not 'killed', because this was a particularly culty and extremely Pythagorean punishment: anybody who broke the rules of the compound, be it the rule against seeing your teacher's face or the rule against putting your left shoe before your right, would die in every way except, you know, the most important one. They'd make you a tombstone and everything.

Other mortal sins – well, mortal-adjacent sins, I guess, or maybe even just a neat way to get a free tombstone on your way out – included dropping crumbs and breaking bread, which sounds harsh, but at least it would have made the crumb thing easier to follow. And as if all that wasn't bad enough, you couldn't even do the decent thing after finishing your sandwich and jump in a grave without unnecessary faff, as it was also verboten to bury the dead in wool.

And all this would be little more than an interesting piece of trivia about community-induced OCD, except that these commandments ended up taking at least two lives thanks to their bananas specificity. And one of those belonged to Pythagoras himself.

So, the thing you have to know about Pythagoras was that he really, really hated beans. Total leguminophobe. And he was evangelical about it. He wouldn't eat beans; his followers were banned from eating beans; heck, there's even a story about him talking an ox out of eating beans. Do you know how determined you have to be to talk a gigantic cow out of eating easily accessible vegetation? Pretty damn determined is the answer.

'But why?' you may ask. 'Why would you hate the humble bean, which never did anything worse than make you fart in front of your elderly grandmother?' Well, it's a good question, and we don't really know the answer. According to Aristotle, the musical fruit was eschewed as a political protest against democracy, while another theory holds that Pythagoras believed farts were your soul escaping, which, as a moral tenet, is almost good enough to justify the whole cult thing.

The most popular view, though, is that the Pythagoreans thought beans were made of the same, well, *stuff* as humans. And you can't fault their logic: have you ever squinted at a bean? It looks a bit like a human foetus, doesn't it? So eating beans is basically the same as cannibalism.

No, seriously, that was their logic.

Anyway, so far so manic-pixie-dream-philosopher – loads of people have weird phobias. I myself once knew someone with a fear of multi-storey car parks. But you'd hope, if yours was an inanimate vegetable, that you might be able to overcome it for a few minutes to literally save your life.

Not so Pythagoras. After pissing off a local nobleman – which you might argue was his first mistake – Pythagoras found himself running from an angry mob who'd decided they'd had quite enough of these fancy-pants philosophers lording it over the common folk. They descended upon the Pythagoreans' hippy commune, torched the buildings, stabbed any cultists they caught running from the flames, and chased the rest into the surrounding countryside.

Unluckily for the mob, who wanted to rid their village of these weirdos as permanently as possible, one of the escapees happened to be Pythagoras himself.

Luckily for the mob, Ancient Greek farmers liked to grow – you guessed it – beans.

Just as it looked like Pythagoras was going to make it out alive, he stopped running. A huge field of beans lay in front of him, halting him in his tracks.

You can see his dilemma: behind him, a great big angry mob set on his brutal murder; in front, some very small vegetables. It's a choice nobody should have to face. But take too long on a decision and the world will often make it for you – and eventually, the townspeople caught up with him.

And so, one of the most famous geniuses in history ended up dying for the sake of some beans: a pretty ridiculous end to a pretty ridiculous guy. But at least *his* demise came at his own stubborn bean-obsessed hand – because there was another famous Pythagorean who wasn't so lucky.

Hippasus of Metapontum may not be a familiar household name, but he should be. The reason why he isn't may well be because he was murdered by a bunch of raving Pythagoreans, which is especially stupid considering the reason they did it, which was, to put it bluntly, because he was too good at maths.

So, I'm sorry, but it's time for a quick number theory lesson – don't worry, it's very light, and I promise* it'll be the only one.

OK: any number can be put into one of two categories. Well, actually, any number can be put into one of about a billion categories, but the two we're concentrating on are the categories marked 'rational' and 'irrational'. In mathematical terms, a rational number is one that can be written as a fraction of integers – that is, whole numbers – and an irrational number is one that, well, can't.

If that doesn't mean much to you, here's a shortcut: irrational numbers are easy to spot, because they're quite often written with symbols instead of numbers. There's a reason for this, which is that irrational numbers are actually impossible to write with numbers. Like, you know those headlines we see every so often about some new supercomputer that's calculated the first fourteen zillion digits of pi? The reason they can do that – and the reason they'll never stop – is because pi is irrational. If you write it out in numbers, it will go on for ever – *literally forever*, not just ages and ages – which is precisely why we just write 'π', if only to save on printing costs.

But because of this ... let's call it *messiness*, there have historically been a lot of people who didn't much like irrational numbers, and the Pythagoreans were no exception. Their very existence seemed to clash with the core Pythagorean philosophies of mathematical harmony and balance, and so, the Pythagoreans concluded, they could not exist.

---

* This is a lie; I love maths too much – sue me.

Now, apart from giving your local neighbourhood mathematician an anger migraine, believing there's no such thing as irrational numbers isn't exactly that terrible. Even rocket scientists only use pi to fifteen decimal places at most; it may pain me to say it, but nothing really *changes* just because you know that a number is irrational. It's basically just a way to say 'we can't write this value down as a fraction, so if we want to use it in real life, we're just going to have to use best estimates instead'. The only difference with the Pythagoreans is that they were saying 'we can't write this value down as a fraction *yet*, so if we want to use it in real life, we're just going to have to use best estimates instead'.

Here's a bit of advice, though: if you're going to deny the existence of irrational numbers, and indeed make it a central tenet of your belief system that the irrational numbers cannot exist, then it's probably a good idea not to *also* teach followers to revere geometrical shapes and pursue beautiful and logical proofs.

Because here's the thing: even the simplest, most kindergarten-level, baby's first shapes are, in mathematical terminology, crammed full of irrational numbers. Circles gave us pi, for example, and squares and triangles reveal the square roots of two and three respectively. Pentagrams, the Pythagoreans' most favoured shape of all, cannot be drawn without including the square root of five. In short, if you're in a maths cult devoted to figuring out the hidden ratios and harmonies of the universe and your special sacred symbols include pentagrams and triangles, it's really only a matter of time until somebody proves that irrational numbers exist.

That person was Hippasus.

We don't know exactly what number it was that Hippasus proved was irrational, but it was probably the square root of either two or five. Most of the details of the discovery have been lost to myth and mystery over the past couple of millennia, and to be honest, it might not have even been Hippasus that did it – but if it wasn't, then we'll never know who it was, so we may as well just label whoever did it

'Hippasus' whether it was him or not. One thing all the stories do agree on, though, is what happened when the news got out.

These days, proving the existence of irrational numbers is a standard first-year university problem for maths students – that is, it's not *easy*, but you're not going to win the Fields Medal* for doing it or anything. But to be the first person *ever* to figure it out? That must have felt amazing. It's no wonder Hippasus wanted to share it with his fellow weirdos: he'd not just come up with a beautiful and unexpected proof, but the result was something so significant that it would likely upend the entire way the Pythagoreans saw the world.

So, I don't know if you've ever witnessed a cultist facing a paradigm shift, but ... well, sometimes it doesn't go so well. Faced with incontrovertible evidence that some numbers could not be expressed in a neat little fractional form, the Pythagoreans reacted in probably the worst way possible, which is to say: they murdered him.

The official story was that Hippasus died as a punishment from the gods for his impiety. What that means in practice is that the Pythagoreans chucked him into the sea and watched him drown, which many people would argue is an overreaction to what is essentially an extra-credit maths problem. But they were safe: the hole in their religion-slash-maths class-slash-murder club foundations had been covered up, and nobody need see or talk about it ever again.

Of course, that couldn't last. The trouble with murdering somebody over a fact that is not only true, but will always be true, has to be true, and anybody with enough of a mathematical bent could figure it out if they happened to be in the right frame of mind one day, is that eventually you're going to look *incredibly* silly. The Pythagoreans may have got away with it for longer than most, but eventually the

---

* There is no Nobel Prize for mathematics. There are two possible explanations you will see for this: the first, funny but almost certainly false, is that Alfred Nobel's wife had an affair with a mathematician, and he was so mad about it that he decided to snub the entire field for the rest of all time. The second possibility is much more boring, but probably true, and it's that he forgot.

knowledge of irrational numbers made its way back into Europe: Indian mathematicians had figured them out way before Hippasus in any case, and within a hundred years or so, other Greeks such as Eudoxus were trying to tackle them again.

They weren't very good at it, I'll grant you – the Ancient Greeks got way too hung up on geometry, and it really held them back in some ways – but then, in around about the eighth century CE, something mathematically miraculous happened: the Islamic Golden Age. Six hundred years of science, maths, art and culture, flourishing like never before – and one of the very first things these scholars discovered was algebra. And you can't do algebra without irrational numbers.

The Islamic world, at least until the eleventh century or so, stretched as far west as Spain, and as such so did their mathematical discoveries. And so, at long last, Europe found out about irrational numbers – not by inventing them themselves, but by nicking them from someone else.*

It's kind of hard to come up with a moral from all this madness, but I guess if I had to come up with something, it would be this: never do your maths homework. You never know when it might get you killed.

Oh – and if you're ever tempted to announce you'd 'rather *die* than even *touch* those beans', remember what happened to Pythagoras. Somebody might just take you up on that offer.

---

* Something which they were no doubt so ashamed of that they never once did it again.

2.

Confucius Was an Ugly Nerd with Low Self-Esteem

On a purely numbers basis, Confucius – or Kong Qiu, to use his given name – was probably the most influential person ever to have lived. His teachings formed the basis of society across vast swathes of East Asia, and were displaced as the main state ideology in his native China only in the twentieth century, 2,400 years after his death.

Despite the wholesale rejection of his philosophy in the decades following the downfall of dynastic China, Confucianism, with its foundational ideas of self-improvement, self-discipline, the cultivation of virtue and the idea that humans are essentially good, still holds strong across the world. There's even a Holy Confucian Church in China now – Confucianism, it turns out, was stronger even than Maoism. His followers celebrate Confucius memorial ceremonies each year; various Eastern religions consider him a prophet or divinely inspired scholar, and as of January 1973 there's even a teeny planet floating somewhere between Mars and Jupiter that bears his name.

If you had told Confucius all that, though, he'd probably have laughed at you, or at the very least scolded you for being impertinent to your elders. That's because Confucius left the world utterly convinced he was a failure: his power and influence were gone, his ideas had been rejected, and his country was about to fall into centuries of civil wars. Of the 3,000 or so students he taught, only seventy-two were said to be true disciples, and it's really thanks to them that we know what we do about the man today.

Confucius lived a true rags to riches – and then back to rags again, kind of – story. He definitely didn't have what you'd call an auspicious start: he was born in around 551 BCE to a mother who was literally a child and a father who was both in his seventies and the kind of guy who thought it was cool to knock up a teenager in his seventies. Like many terrible people throughout history, his dad was convinced that women and people with disabilities weren't worth shit, and so despite already having nine daughters and one son with a club foot, he saw Confucius's mum as his last chance to create his legacy.

Confucius's father, according to legend, was a fierce warrior in his day, although to be fair he had an advantage in this regard due to being frankly enormous. He was also butt-ugly – and unfortunately for little baby Confucius, he apparently took after his daddy in this regard.* He was born with 'the lips of a cow and paws of a tiger, the shoulders of a mandarin drake, and the spine of a tortoise. He had a wide, open mouth and a long neck, and his forehead was shaped like an inverted roof.'† Confucius also had a crooked nose and a big bulbous head, he would grow up to be buck-toothed and awkward, and just in case he ever started feeling like anything more than a misshapen weirdo, his parents decided to actually name him Qiu, or 'mound', after the odd shape of his skull.

---

* According to some sources, in both regards: one ancient source records the adult Confucius as being nine feet and six inches tall.

† That's how he was described by the Ming dynasty historian Feng Meng-Lung, who admittedly *was* writing a couple of millennia after Confucius was born, but what are you gonna do? – it's the best we've got.

As septuagenarians were wont to do back in those days, his dad died pretty quickly after Confucius was born, and good riddance, really, because who needs that kind of negativity in their life. The downside, though, was that his family shunned Confucius and his mum for the rest of their lives, and the pair were left to fend for themselves in Iron Age China.

Confucius's first teacher was his mother, but at the age of seven he started attending a local school. He fricking loved it. He always would – he would later summarise his life as:

> At fifteen, I had my mind bent on learning.
> At thirty, I stood firm.
> At forty, had no doubts.
> At fifty, I knew the decrees of heaven.
> At sixty, my ear was an obedient organ for the reception of truth.
> At seventy, I could follow what my heart desired without transgressing what was right.

Unfortunately, as anyone who's ever been a child at school will know, being smart, ugly and poor rarely goes hand in hand with being popular. The young Confucius seems to have been great at mastering the 'Six Arts' required by an ancient Chinese education – rites, music, archery, chariotry, calligraphy and mathematics – but he was a bit of a social outcast, and instead of playing with the other children, he spent his time re-enacting ancient rituals and reading history books.* Still, if his particular combination of looks and charm hadn't made him a hit with the other kids, at least he wouldn't have to put up with their bullying for long: in the tradition of poor kids throughout history, he left school and started taking a bunch of random menial jobs to help support his mum.

---

* Or, at least, the ancient Chinese equivalent of history books.

So, Confucius hasn't had a great time of it so far. He's grown up an ugly, poor nerd with no dad and a mum who's practically a child herself, and now he's taking jobs like sweeping streets and herding cows just to pay the bills. But this combination of apparently bad luck is exactly what turns him into the semi-mythical figure he is today – even if he never thought of himself as anything more than a fairly rubbish teacher.

See, working all those crappy jobs gave Confucius something not many people had back then: perspective. China was basically feudal at the time, and just like in some young adult dystopian novel, non-noble society was strictly divided into four or five 'occupations', arranged in order of their supposed worth. At the top they had the scholars; then came farmers, artisans, and finally merchants, who it seems very rarely catch a break when it comes to social standing. Then, below even the merchants, you had those who existed outside of the standard classifications completely: the sex workers, slaves and musicians.

Now, these occupations weren't hereditary – the son of a farmer could become a scholar, for instance – but it was believed that if they were allowed to mingle, then people would get confused and forget how to work or speak and, before you knew it, society would break down completely. A wise ruler, philosophers of the time advised, would make the various peoples live apart, work apart, and never mix with each other, and what this meant in practice was that all the scholars of the time were writing theories about how to be a good shoemaker or whatever without ever even talking to a real-life craftsman.

Confucius, though, approached life differently. His mum died when he was seventeen, and once he had recovered from the grief of losing literally the only constant in his life, he finally experienced a stroke of good luck: a local lord made him overseer of a granary. This was a cushy appointment, particularly given where Confucius had started out, and if he had played his cards right he could have stayed comfortably in this new occupation for the rest of his life. Instead, he became a teacher.

And Confucius as a teacher was a bit ... weird. He was super forward-thinking and progressive, while at the same time being extremely conservative and ritualistic. That sounds like an obvious contradiction, and it is, but from his point of view it made complete sense. He had just one objective, and it was as simple as it was laughably ambitious: China, he thought, was going to the dogs, and he wanted to save it.

This is where all those years of being rejected by his schoolmates finally came in handy. Confucius looked at the world around him and saw chaos – but in the China in those history books he'd spent all his playtimes reading alone in the cafeteria, everything had been wonderful. Confucius figured that something must have gone wrong at some point, and if he wanted to fix society, all he had to do was figure out what had worked once and then get people to do it again.

Even though today we think of Confucius as a philosopher, it's important to remember that he's very different from Western philosophers like Aristotle or Plato. He wasn't interested in the kind of head-in-the-clouds metaphysical questions that they spent their time arguing over; he wanted practical, practicable solutions for what he saw as the problems of life. He didn't write abstract discourses or logical treatises like Western philosophers – he taught in person, by talking to people, and never expected his lessons to be written down at all. And he didn't see his job as coming up with new ideas, but rediscovering old ones – which meant he basically had no choice but to be *extremely* conservative. He was obsessive about tradition and ritual, and honestly he must have been pretty unbearable to live with: according to the *Analects*, the collected sayings and teachings of Confucius compiled by his disciples after he died, he lived by such rules as refusing to wear red or purple, refusing to eat meat that hadn't been cut the way he liked, refusing to eat meals served with the wrong sauce, refusing to drink wine or eat meat bought from the market, refusing to sit on a mat that wasn't positioned at the correct angle ... you get the picture. And if he was a pain to live *with*, then pity the poor citizens who had to live *under* him: when he was around fifty, he temporarily got some high-ranking job that let him institute laws in his province like 'men and women must walk on different sides of the street' and 'inventing unusual clothing deserves the death penalty'.

It may seem odd, then, that he was also known for some of the most radical ideas known anywhere at the time. He was willing to teach anyone, no matter their class or occupation – all he asked was that they would follow the doctrine of hard work, study and self-improvement. He thought that education could transcend class boundaries – an idea that certain other nations still struggle with today – and most of his students actually ended up coming from the lower rungs of the social ladder.

Maybe it was because he was so ready to teach anybody regardless of class, or maybe it was because the people really liked his obsession with how straight his mat should be, but either way, Confucius started getting popular. He held small, informal classes in which he trained his acolytes to be good, well-rounded government officials, always with the end goal of influencing the country's Imperial rulers and bringing China back to stability in mind.

And what was his big idea to save the nation? What did all that philosophy really boil down to? In short: everything will be all right if you just set a good example.

It sounds kind of silly, right? I mean, here you are, in 500 BCE China, your dad's some millet farmer, you live in a feudal state run by a guy who could be called a duke or a warlord with pretty much equal accuracy, every aspect of society feels chaotic and fragile, and here comes some buck-toothed big-headed dude saying all this can be fixed if we all just … behave ourselves? I mean, what's his solution for homelessness – 'just buy a house'? Thanks, Confucius.

He really believed it, though. When one of his students asked him how to run a government, his answer was simple: 'Encourage the people to work hard by working hard yourself.' When the student followed up with the ancient Chinese equivalent of 'um, could you explain more please that's pretty vague', Confucius simply replied, 'Do not allow your efforts to slacken.'

'If a ruler himself is upright, then all will go well even if he does not give orders,' he taught, 'but if he himself is not upright, then even if he gives orders they will not be obeyed.'

But there's only so long you can be a popular and successful teacher devoted to reforming society in a country run by a pack of warring feudal lords before somebody decides you're a problem, and Confucius's time ran out right around the time he was sentencing people to death for wearing zany hats. Legend has it that his political rivals sent his boss a 'gift' of eighty young, beautiful girls – that's eight-zero human beings – who so diverted his attentions that he was mysteriously compelled to fire Confucius.

Humiliated, Confucius went on tour. He became a wandering teacher, collecting new pupils and meeting other superstar philosophers like Lao Tzu, the founder of Daoism. He doesn't seem to have reflected much on his manners during this time: at one point, the story goes, he met a woman whose husband and child had been eaten by tigers, and his reaction was to tell her that 'an oppressive government is more terrible than tigers', which I'm sure she found very comforting.

Eventually, after thirteen years, Confucius went home, hoping once again to get enough influence to shape the direction of Imperial China. But whatever those eighty girls had done, it had stuck, and Confucius never again found himself in a seat of power. Neither did any of his students.

Now an old man and convinced that all his teachings had been for nothing, Confucius fell into a depression that lasted the rest of his life. He spent his last years reading and commenting on the same history books that he had first read back in school, and when he died aged seventy-two* he likely thought nobody would ever know his name

---

* Take this with a grain of salt; seventy-two is an 'auspicious' number in Chinese tradition, so his age could have been fudged a little.

again. One of his last known lessons, said a week before he died, was a lament that 'The great mountain must crumble. The strong beam must break. The wise man must wither away like a plant.'

But despite all his flaws – and counter to his own perception of himself as inconsequential – Confucius was to become a legend. Nearly three centuries after he died, his philosophy was made the official ideology of China – a status it would enjoy for the next two millennia or so. He was, by his own admission, not an original thinker or a particularly special human specimen, but his ideas are still popular today because he tapped into something fundamental about the human experience: the desire to simply be a better person today than you were yesterday.

Confucius thought that no matter who you were, you could always strive to learn, grow and improve yourself – in fact, more than that, he saw it as your moral duty to do so. But most of all, he believed that anybody – rich or poor, high-born or low – had the ability and responsibility to act virtuously, contribute to the well-running of society, and keep their sauces in order. For the first time, morality and ethics became something that wasn't delivered by some king or duke, but accessible within everyone – whether they be the Emperor of all China, a lowly artisan, or even an awkward, bulbous-headed, buck-toothed old man convinced his teachings had, in the end, been completely worthless.

3.

Never, Ever Hire Leonardo da Vinci

If Leonardo da Vinci were alive today, there'd probably be a TV show about him. But not the kind you're imagining, where they take some amazing shining light of culture and tell the world about their work and life or something. It would be more like one of those shows where some TV presenter LARPing as a cop sets up a sting in a dingy rented flat, and it would be called something like *Catching a Rogue Trader*. And you know what? They'd have him bang to rights.

Leonardo da Vinci, so named because he was called Leonardo and he came from the city of Vinci, is rightly regarded today as one of the greatest geniuses who ever lived. But that great big brain came with some drawbacks. For Leonardo, it was the fact that nothing could hold his attention for any longer than about twelve seconds. For everybody else, it was ... well, also that. As the Renaissance biographer and artist Giorgio Vasari wrote: 'Truly marvellous and celestial was Leonardo ... and in learning and in the rudiments of letters he would have made great proficience, if he had not been so variable and unstable, for he set himself to learn many things, and then, after having begun them, abandoned them.'

He seems to have been that kid who teachers fear: relentlessly interested in absolutely everything, and so bright that before you're done introducing them to a topic they're already bombarding you with questions about it that you can barely understand, let alone answer. He was into art, maths, music, engineering, sculpture, you name it, and that boundless enthusiasm came out in doodles and drawings and sketches of anything that happened to cross his mind, which is to say, everything.

By the time he was fourteen, his dad had evidently picked up on the fact that he needed an outlet for all that energy before he imploded or something, and he set Leo up with a job as a *garzone* – a studio assistant – for his artist friend Andrea del Verrocchio. This would prove to be either the best or the worst decision Verrocchio ever made, depending on how you look at it: on the one hand, he now had in his charge a brilliant young student to help him sculpt and paint better than he ever had before, but on the other, Leo apparently became so good that by the age of about twenty his addition of a little angel in del Verrocchio's *The Baptism of Christ* was so much better than his master's that del Verrocchio was moved to give up painting for ever.*

Leonardo would stay with del Verrocchio for a total of ten years, meeting fellow future household names (and ninja turtles) and perfecting the art of ... well, art, until eventually, in his early twenties, he set out on his own. So, not long after he ruined his master's career by being too good at painting angels, he received his first commission: a picture of the Adoration of the Shepherds† for the chapel of San Bernardo in the Piazza della Signoria, Florence.

---

* This story, originally recorded by Vasari, is generally believed to be apocryphal, but it's a good 'un nonetheless. It is certainly true, though, that Leonardo collaborated with del Verrocchio on certain paintings – it was pretty standard practice back then. Leonardo also modelled for del Verrocchio on a few occasions, so if you want to see what a young Leo looked like, check out del Verrocchio's statue of David in the Bargello Museum, Florence.

† The bit in the Bible where the shepherds come and coo at the baby Jesus.

And this is where Leonardo, now free from the shackles of a boss who knew what a contract was, started becoming something of a menace. He was paid an advance for the piece, but only got as far as making a mock-up – after that he gave up, and never completed the commission.

Lucky for him they didn't have Yelp back then, because the following year he was commissioned again, this time by the monks of San Donato a Scopeto. They wanted him to paint the Adoration of the Magi.*

Again, Leonardo started the piece, and again, he never finished it. He got further than with the shepherds, but to this day you can see in the centre of the exquisitely composed scene† that Mary, Jesus and quite a few Magi are all just kinda … sketched in. And it's not like he ran out of time: he had been given a schedule of more than a year to complete the project, but he abandoned it after just seven months.

But why did Leonardo never get round to colouring in the main characters in his painting? Well, it seems it was around this time he entered what we'd now recognise as his emo edgelord phase. Instead of painting cute cherubs and baby Jesuses for the rest of his life, he decided to write to the Duke of Milan, Ludovico Sforza, offering his services as a military engineer and inventor – who could also paint, by the way. He sent the duke sketches of war chariots with scythed wheels, and UFO-looking tanks propelled by a crankshaft, and a huge goddamn crossbow that would need a whole group of people to fire it. He made himself a solid silver lyre – a stringed instrument kind of like a medieval cross between a guitar and violin – in the shape of a horse's skull, which is undeniably pretty metal. And to top it all off, he accepted a commission from Ludovico that both of them were excited about: a great big sod-off horse, sculpted entirely out of pure bronze.

---

* See previous footnote, but replace 'shepherds' with 'wise men'.

† In which Leonardo himself once again makes an appearance: if you ever see it, look at the guy on the right-hand side of the crowd – it's generally thought to be a self-portrait of Leonardo.

Now, if you go to Milan today, you will indeed see a huge bronze statue of a horse standing, mid-trot, on a plinth in the Hippodrome de San Siro. It's there. It's finished, too – no missing hooves or a tail that's only sketched in or anything. But it took ... a little longer than either Leonardo or Ludovico expected. See, Leonardo wanted everything about it to be perfect, and he meticulously set about researching the anatomy of horses – he even wrote a treatise on it. He also figured out a plan for casting the statue, and wrote a treatise on that. He had everything ready to go. And then he ... didn't.

Instead, he took up another painting commission, from the Confraternity of the Immaculate Conception in Milan. He was to create a life-size portrait of the Virgin Mary, the baby Jesus, John the Baptist (also a baby), and an angel, Uriel, all having a little get-together in a rocky outcrop. We know it today, for obvious reasons, as *The Virgin of the Rocks*.

Well, to be specific, we know it as the *Louvre Virgin of the Rocks*. That's because there's actually two of them: one – the original – is in the Louvre, and the other – an almost identical copy – is in the National Gallery in London. And the story of why these two twin Virgins exist is yet another tale of Leonardo being basically a bit of a wide boy: upon the piece's commission, Leo was paid 100 lire, with the promise of a monthly stipend of forty lire for the next couple of years. He was also set to receive a lump sum when the piece was finished, which everybody agreed would be by early December 1483, in time for the feast of the immaculate conception.

Oh, those sweet innocent monks. Probably the first sign that things weren't on track would have been when the deadline passed with no finished painting forthcoming, but it wasn't until the best part of a decade had passed that things got properly hairy. Leonardo, along with another guy who was working on the piece, wrote to the Confraternity demanding an extra 1,200 lire, which, you might notice, was more than the entire fee they had agreed in advance. Costs had run over, they said. The monks countered with 100 lire, to which Leonardo replied along the lines of 'screw that, mate' and literally just took the painting back.

And sold it. To someone else.

Now, maybe he had a crisis of conscience, maybe he was just aware of the kind of trouble a person could get into on the wrong side of the Catholic Church in Renaissance Italy, but this is when Leonardo started painting his replacement Virgin. And it might all have ended amicably, except that it took so long to complete that the King of France, the ironically named Charles the Affable, invaded Milan, prompting Leo to flee the city in 1499 with yet another commission unfulfilled.

So now these monks are down one and a half paintings *and* nearly a thousand lire, and so they did what any of us would do when faced with a contractor who refuses to deliver: they brought legal action. It took another decade, but Leonardo was eventually forced to go back and finish the piece, which was finally installed in August 1508, a mere twenty-four years and nine months after the original commission deadline.

Still, at least all that time spent stringing along the monks would have given him time to sculpt that horse, right? And so, in 1490, eight years after he first moved to Milan with naught but a couple of lire and a dream of a gigantic bronze pony, he wrote in his diary: 'Time to make a fresh start with the horse.'

Look, it's not like he'd made *no* progress: by 1493, he had a complete model of the horse sculpted out of clay. He was, he said, ready to start casting the bronze. There was just one problem: the aforementioned French invasion. This time, it genuinely wasn't Leonardo's fault that his work got held up – Sforza gave the seventy tons of bronze that had been intended for equine posterity to his father-in-law to make cannons. Cannons which apparently didn't work very well, by the way, because the French did in fact successfully take over Milan, and the clay model was put to use by their archers for target practice.

So, since he literally couldn't make the horse for now, he took another commission: a fresco, for the walls of the convent of Santa Maria delle Grazie, of the Last Supper. Yes, *that* Last Supper.

Now, to give Leonardo his dues here, *The Last Supper* is, in fact, finished.* But that doesn't mean his bosses were happy with his work: firstly, the scene he set out would have been unlike any Last Supper they'd seen before – they were probably expecting a big old Jesus at the top, and a bunch of little disciples at the bottom, all calm and serene as they took Communion. That's what Last Suppers looked like, everybody knew it, so what was this upstart from Vinci doing drawing Jesus the same size as his disciples, who by the way are all aghast and grumpy and *all over the fricking place might I mention*. It would have been shocking.

And even worse, the fresco was taking *years* to complete – so long that the convent's prior started complaining. He called Leonardo lazy, saying that he was spending all his time wandering around town instead of working. Leonardo took this well: he wrote to the head of the convent, the prior's boss, saying that 'only the head of Judas remains ... and he was an egregious villain. To this end, for about a year, I have been going every day to the area where the ruffians live, but have not been able to find a face corresponding with what I have in mind. But I may take the features of the prior who came to complain, who would fit the requirements perfectly!'

Then, in 1500, the French finally finished their invasion of Milan, and Leonardo once again fled – this time to Venice, where he briefly worked as a military engineer, designing among other things an absolutely terrifying elephant-cosplay looking 'diving suit', before quickly heading back to his hometown of Florence. For a while in 1502 he

---

* Although if you've ever seen it in person, you'll know there's a bloody great bricked-up door underneath which cuts into the middle of it, chopping off Jesus's feet, among other things. Apparently it was put in so the monks and nuns could get to the refectory quicker, which just goes to show that even the Lord and Saviour Himself is no match for the divine power of pizza.

entered the service of Cesare Borgia, a man so ruthless that he literally became the inspiration for Niccolò Machiavelli's *The Prince*, basically a handbook on how to be a bastard.* According to the Venetian ambassador, Cesare's Rome saw 'every night four or five murdered men ... discovered', and that may be why Leonardo didn't stay very long. In 1503, he moved back home again, and within a few months he had picked up a new commission: a portrait of a local noblewoman named Lisa del Giocondo. We know her today with a slightly more formal title – something like 'Madame Lisa'. Or, to use the Italian term: *Mona Lisa*.

But while he was painting what we now know as his defining work – if not *the* defining work of all Western art – he was also busy with another piece: a commission for a mural of the 1440 Battle of Anghiari, to be displayed in the town hall of Florence. It was sort of a grudge match, actually: the opposite wall was going to have a mural too, but painted by Michelangelo.

Now, it's tempting to think, what with them being in the same place at the same time, doing the same thing and probably hanging out at the same bars, that Leonardo and Michelangelo must have been friends. They weren't. As far as Leonardo was concerned, Michelangelo was a disturbingly religious and sexually repressed posh boy; meanwhile, to Michelangelo, Leonardo was a grubby old dandy who never bothered to finish anything.† Which was true, but

---

* This is not exactly fair to Machiavelli, but the treatise *is* explicitly amoral. In any case, there are far more interesting things about Cesare to talk about, like the fact that he was somehow the son of a Pope, is to this day the only person to ever quit being a cardinal – apparently he just wasn't that keen on the whole 'God' aspect – and the time he supposedly set up a party with the Pope involving 'fifty honest prostitutes' (that's genuinely the term used) being made to crawl around on their hands and knees, naked, picking up chestnuts off the floor, before the whole thing devolved into a gigantic orgy.

† There's a very famous story where Leonardo is asked to explain a section of Dante to a couple of guys in the town square, when who should walk by but Michelangelo. 'Michelangelo will be able to explain it for you,' says Leonardo, to which Michelangelo replies, 'No, you explain – you who have undertaken the design of a horse to be cast in bronze but were unable to cast it, and were forced to give up in shame.' Meow.

to be fair, neither did Michelangelo, and the mural competition would eventually come to a very anticlimactic end when both artists just kind of got distracted by other things and never returned to it.

For Leonardo, those other things were the sciences. His notebooks from this time were already filled with anatomical drawings way past what any of his contemporaries were capable of – mostly because, unlike them, he was happy to chop up corpses for reference material. He had a real passion for drawing what he called the *figura istrumentale dell' omo*, which is an extremely fancy way of saying dick pics, and he figured out things like gravity, and sound waves, and described turbulence in a way that's strikingly similar to what you'll find written in a modern physics textbook. He even made strides in geology, proving that the Earth's crust has moved over millions of years and throwing 'known' facts like the Biblical flood into doubt.

And all of that is really, really impressive, except for two things: one, the fact that he never showed anybody else, so all that scientific progress was locked up in his notebooks instead of changing the world, and two, he still hadn't delivered on that bloody horse.

By 1512, now nearly three decades since he first pitched the horse, Leonardo was back in Milan and once again working on designs for the statue. Within the year, he had once again abandoned it, as Milan was invaded *again* – not by the French this time, but by everybody else trying to take it *from* the French. Leonardo fled *again*, this time to the safety of the Vatican, which had just come under the rule of Pope Leo X.

Now, Leo X liked Leo dV, and he commissioned Leonardo to compose a painting for him. And you'd think, given that it was the actual Pope, that Leonardo would finally take this assignment seriously, right?

So imagine Leo X's surprise when, seven months after ordering the painting, he asked Leonardo how it was coming along only to receive the reply that it hadn't been started yet, due to Leonardo getting carried away designing a new type of varnish.

'This man will never accomplish anything!' the Pope reportedly prophesied, upon learning this. 'He thinks of the end before the beginning!'

In fairness, Leonardo knew he had upset his patron, and he tried to make it up to him in classic Leonardo style: by presenting him with sketches and a treatise on vocal cords. Which is sweet, but probably not what the Pope was after when he commissioned a religious painting.

So Leonardo moved on again, leaving yet more unfinished work in his wake. He moved to France, where he gained the favour of the king – a new guy who was appropriately named Francis of France. He only completed one more painting before his death – a kind of horny portrait of John the Baptist, modelled after his student and possible boyfriend Andrea Salaì – but he did also build a working mechanical lion for a pageant, which walked to the king and opened its chest to reveal a bouquet of lilies when tapped with a wand, which is worth at least two or three paintings he'd never finish anyway.

Leonardo died in 1519, in France, aged sixty-seven, leaving the world almost as many abandoned, half-finished, or outright lost masterpieces as complete ones. He had filled notebook after notebook with scientific discoveries that would take the rest of the world centuries to catch up with, but which no one saw, and we're meant to look at that as a sign of genius rather than short-sightedness.

Even his most famous work, the *Mona Lisa*, he never finished – he took it with him to France and insisted until his death that there was still more to do before it was complete. Which kind of puts the painting in a new perspective, really – the most famous image in the world, and yet nobody *really* knows what it was meant to look like finished.

And as for the horse – well, that finally went up in 1999, thanks to a sculptor named Nina Akamu. Like Leonardo himself, she put painstaking research into the statue, reading Leonardo's notes on anatomy and philosophy, learning about horse breeds in fifteenth-century

Italy, even looking into the teachers whose influence can be seen in Leonardo's own work.

And then she did something Leonardo, arguably history's greatest genius, never could.

She built the damn horse.

4.

Galileo Utterly Fails to Read the Room

Imagine you wake up one day to an alert on the news – it's Professor Brian Cox, and he's making the headlines with a new theory. According to him, we've been wrong about light. Forget all that stuff about rays of electromagnetic radiation reflecting off things and brightening up the darkness – it's actually the exact opposite, according to the Professor. Turns out, light is everywhere, all the time, and it's only when it's broken up by rays of darkness, emitted by physical objects, that we see things like shadows or the darkness of space.

Not only can he replicate all known physical theories with this new model, it also answers a few questions. Olbers' paradox, for instance, which asks how, with the unimaginable number of stars 'emitting light' in the universe, the night sky can possibly be dark. Well, now we know, says Cox: there are more planets than stars in space, and they emit the darkness that makes up most of the night sky. It's only in the rare spots where the dark rays don't reach that we see what we think of as a 'star'.

Now, Brian Cox is a pretty well-known and impressive intellect, but even so, this is a bit far-fetched, you think. We've known how light works for centuries, we don't need some new explanation – especially not one that is quite clearly, even to an idiot like you, arse-backwards.

That must have been what it felt like when, in 1543, the seventy-year-old Nicolaus Copernicus published his magnum opus: *De revolutionibus orbium coelestium*, or 'On the Revolutions of the Heavenly Spheres'. For the first time in human memory,* somebody was suggesting the unthinkable: that it was the Earth that moved through the heavens, and not the Sun. Even before it was published, it had been denounced by figures like Martin Luther, who called Copernicus 'an upstart astrologer' and a 'fool'.

'[He] wishes to reverse the entire science of astronomy;' Luther wrote, 'but sacred Scripture tells us … that Joshua commanded the Sun to stand still, and not the Earth.'

No wonder, then, that Copernicus waited until the very last minute to publish his theory of heliocentrism, dying just a few months after he delivered his manuscript to the printers. It wasn't quite as controversial as you might think, though, for a couple of reasons: first, because Copernicus was careful to write it in such highfalutin technical language that almost nobody could actually understand it, and second, because a passing Lutheran theologian named Osiander unilaterally added a bunch of disclaimers to the text saying things like 'oh, by the way guys, this isn't real, it's not science it's just a fun what-if, I promise.'

This would prove to be one of those tiny moments that changed the world. The guy who was supposed to be in charge of printing the manuscript was Georg Joachim Rheticus, a mathematician and general polymath who was Copernicus's only student. But just before the book reached him in Nuremburg, he left for a new job in a different city, and Osiander took over. His decision to add a preface

---

* Though it should be noted, not the first time in human history. There were quite a few Ancient Greeks who had figured out the heliocentric model – among them none other than our own Pythagoreans – and medieval Islamic scholars started questioning the Earth-centric model of the solar system in around the tenth or eleventh century.

 Meanwhile, Indian astronomers had them both beat by at least a century or so. 'The Sun causes day and night on the Earth because of revolution,' reads the Aitareya Brahmana, an ancient Vedic text generally thought to date to around the sixth century BCE. 'When there is night here, it is day on the other side; the Sun does not really rise or sink.'

saying the book wasn't intended as truth has split commentators ever since it happened. Some people were horrified that he would take it upon himself to doctor and dilute Copernicus's work like this. Others, however, thought it was a blessing in disguise that avoided the charge of blasphemy on a technicality – and perhaps they were right, since it took the Vatican a whole seventy years before they got around to banning the book. But by that point, Copernicus was long dead, and his mantle had been taken up by a middle-aged Italian astronomer with a charmingly self-referential name.

Galileo Galilei was born in 1564, only a couple of months before Shakespeare, and his parents were clearly fucking sick of him before he even reached double figures. When he was eight, his mum and dad moved from Pisa to Florence without him; he reunited with them a couple of years later only for them to send him off to live in a monastery more than twenty miles away. His father did eventually summon him back home, but not because they missed him or anything; he was simply enjoying himself too much at the monastery.*

As you might expect from a music teacher with six children to support, Galilei Sr was painfully aware of the importance of money, and he had already decided long ago that no son of *his* was going to be taking vows of poverty. Galileo would be educated at home in Florence until he was old enough to go to university and enrol in medical school, and that would have been that – had he not accidentally run into that most beguiling of mistresses: mathematics.

Rather than 'Intro to Leeches' or whatever they taught back in the day, Galileo attended lectures on things like geometry and natural philosophy; rather than bringing home a pile of laundry and a few latent venereal diseases for the summer holidays like a modern university student, Galileo brought his maths lecturer. Finally convinced that his son had no interest in becoming a physician, Galileo's father relented, and allowed his son to study mathematics alongside his medical studies.

---

* Not like that. Apparently the young Galileo unironically enjoyed living far away from the rest of the world under the strict rules of the Camaldolese Order.

Galileo Utterly Fails to Read the Room • 39

If he was hoping this would act as damage limitation, Galilei Sr would be sorely disappointed, because by 1585 Galileo had dropped out of uni completely and moved back home to Florence. He spent the next few years mooching around Italy teaching mathematics, which at least meant he probably had a lower kill count than if he had become a doctor, and in less than two years he had put out his first book: *La Bilancetta*, or 'The Little Balance'. It's mostly been forgotten now, partly on account of all the other things Galileo ended up being famous for, but also because on the face of it, it didn't really add anything new to the scientific literature. It was basically a manual for weighing and measuring things, which, frankly, people had been doing for ages by this point, and it had been written by some college dropout nobody who hadn't even bothered to use Latin.

Still, it was a start, and it was enough, in Galileo's opinion, to get him a professorship at the extremely prestigious University of Bologna, still the oldest in the world. Unfortunately, the university disagreed, which can be a risk when you just rock up at a five-centuries-old institute of higher education as a medical school dropout and demand a job as a maths professor. He certainly made an impression, however, since a few years later he was invited to the prestigious Academy in Florence to deliver a lecture on the delightfully medieval subject of the exact dimensions and location of Hell.*

---

* Not as impossible as it sounds. The general shape and geography of Hell had been known for a few centuries already thanks to Dante, who went into some detail in *Inferno* – and, sure, you might be thinking 'But that's just a poem! A work of fiction!' and you'd be right, but that didn't stop the Church from declaring it canon.

Now, *Inferno* is thirty-three cantos long, but the important bits from a mathematical perspective are the following: 1. Hell consists of nine concentric levels descending into the Earth in decreasing order of size – imagine a gigantic, cursed ice cream cone of damnation, basically; 2. It is centred on Jerusalem; and 3. The diameter of the mouth of Hell is as wide as the radius of the Earth. Using Galileo's approximation of that length – 3,245 miles – that makes Hell about 9,587,937,046 cubic miles in volume; using the *actual* radius of the Earth, which is about 714 miles larger, gives an *Inferno* volume of 17,411,555,667 cubic miles.

Now, there are many descriptors that may come to mind when you hear the words 'maths lecture', but I'm guessing 'sassy' probably isn't one of them. Back in the sixteenth century, though, academia was a bit more cut-throat, and seminars could often be ... well, less like you're imagining them today, and more like an epic rap battle in which the goal is to smack down your opponents with the sickest burn possible. This wasn't just a quirk of the time – a renaissance mathematician's entire livelihood could depend on their ability to come up with a devastating geometry-based put-down. In Bologna, there was even a university rule requiring every lecturer to take part in an academic debate at the Piazza Santo Stefano at least twice a year – displays that were nominally an academic exercise, but usually ended up devolving into something of an in-person flame war. Often, they were so popular that there wasn't enough room to house all the spectators.

Mathematics in particular had a long tradition of these 'duels' – and they really *were* duels, with one mathematician laying down the 'challenge gauntlet' (that is, sending their opponent a list of problems to complete within a set time period, and yes it was really called that), which their rival was obliged to reply to in kind. The winner of the duel – the mathematician who answered the most questions correctly – could look forward to fame, fortune, and all kinds of professional rewards; the loser, meanwhile, would lose face and quite possibly their current and future job opportunities. The tradition had been known at least as far back as the tenth century; they were how Leonardo of Pisa – you might know him as Fibonacci – made his name, and they played a crucial part in how the world was bequeathed imaginary numbers.*

---

* That is, numbers which are the square root of a negative number. Technically, it's the story of cubic equations rather than imaginary numbers, but one ended up being necessary for the other; see cubic equations – that is, equations that look like this:

$$ax^3 + bx^2 + cx + d = 0$$

where *a*, *b*, *c* and *d* are numbers that you're given, and *x* is the number you have to figure out – had puzzled mathematicians for ages. Like, we're talking millennia. They definitely *had* solutions – at least sometimes. The problem was, nobody knew how to *get* them. And academia being what it was at the time, there was basically a kind of pan-European arms race going on among mathematicians, all of whom wanted to be the first to figure it out and be able to deploy a brand-new KO move in the next round of maths duels.

Eventually, in about 1510, a mathematician named Scipione del Ferro somehow figured out a way to solve one special type of cubic equation – and then, obviously, since he didn't want anybody to take his livelihood, he kept it completely secret for the next decade and a half, revealing it only on his deathbed to just one person: his assistant, Antonio Fior. Now, Fior was not an amazing mathematician, but he apparently *was* an insufferable braggart, and he was all too happy to let people know that he knew how to solve cubic equations. Unfortunately, he fell into the trap of forgetting that just because you *say* you're good at something doesn't mean you actually *are*, and he started to do very silly things like challenge to a duel one of the foremost mathematicians of the day, a man known to history books for self-evident if slightly mean reasons as Tartaglia (Italian for 'Stutterer').

As far as we know, Tartaglia didn't know how to solve depressed cubics, but he figured if this chancer from Bologna could do it then it couldn't be that hard, all things considered, so he set about finding a way to do it. He was successful: when Fior delivered his thirty problems, Tartaglia managed to solve them all in a matter of hours, while Fior didn't manage to finish a single one of Tartaglia's.

Long story short, Tartaglia became the most famous mathematician around off the back of this, and eventually his method made it into *Ars Magna*, a ground-breaking book on algebra by Gerolamo Cardano that is now considered one of the greatest scientific treatises of the Renaissance. There's just one problem: sometimes there would be a step in the method where you had to take the square root of a negative number.

Of course, anybody who's done primary school maths (but, and this is important, not more advanced maths) knows that is impossible, but Cardano had a sneaky and, for anybody who knows any professional mathematicians, incredibly on-brand solution. Turns out, if you just kinda pretend everything is fine and carry on regardless, the problematic bits often cancel out, leaving you with a perfectly sensible solution – and in fact, Tartaglia ended up losing *another* maths duel later on to a student of Cardano's who had taken advantage of this.

Cardano wouldn't have known it at the time, but his 'fuck around and find out' approach had done more than just give him the edge in any future epic math battles – it had fundamentally changed how we think about mathematics. For the first time, numbers didn't have to mean anything geometrically – after all, how do you measure negative area? – and it was this split that eventually allowed later mathematicians to accept imaginary numbers as a legitimate mathematical tool.

Galileo excelled at this style of academia, which is a polite way of saying he could be a snarky little bitch, and the crowds loved him for it. His lecture on Dante didn't include much in the way of original mathematical work – instead, he basically spent his time at the lectern lampooning another mathematician, Alessandro Vellutello, who had committed the crime of obtaining a result that disagreed with someone from Florence.

We know now – and so did he, just a few years later – that Galileo's arguments in this lecture were completely wrong, but he was nevertheless made Professor of Mathematics at Pisa just a few months later, which just serves to underline how important being able to deliver a fatal one-liner was to the discipline of geometry at this point.

For the next two decades or so, Galileo had it made. After a few years at Pisa, he moved to the University of Padua and a salary three times what he had been earning before. His job was mostly to teach basic geometry and astronomy to undergraduate medical students,* which is a schedule any lecturer today would probably literally kill for; he met Marina Gamba, a woman he fathered three children with but did not marry, and he even designed and built a brand new super-accurate (for the time, at least) telescope – an invention that earned him a whole lot of money *and* made him the guy who discovered the rings of Saturn and the four largest moons of Jupiter. Capitalising on his newfound fame, he moved jobs once again, becoming Chief Mathematician at the University of Pisa – with no teaching duties at all now – and Mathematician and Philosopher to the Grand Duke of Tuscany, which is the kind of fancy title and working conditions most working academics couldn't hope to get even if they *did* kill for it.

All he had to do was keep his smart-arse mouth shut, and everything would be fine.

---

* If you're wondering why a medical student would need to know about geometry and astronomy, it was of course so that they could tailor their diagnoses to their patients' horoscopes. Because, you know ... the past.

To be fair to Galileo, he did try to keep his nose clean. And it can't have been easy: with his new telescope, he was discovering more and more in the universe that, frankly, didn't make any damn sense if the science he had been taught was true. For instance, he could *see* that Venus seemed to have phases, like the Moon, and that meant that it must orbit the Sun. But everybody *knew* that Venus – all the planets, in fact, as well as the Sun and Moon – orbited Earth. Not only did it *literally say so in the Bible* – and that should be evidence enough for your average sixteenth-century scholar with an interest in not receiving a surprise visit from the Inquisition – but rational scientific enquiry supported it too: how, for example, could stars be visible to the naked eye if Copernicus was right and they were all hundreds of times further away than the Sun?* How could a reasonable person, as Tycho Brahe put it, ascribe 'to the Earth, that hulking, lazy body, unfit for motion, a motion as quick as that of the aethereal torches'? No, clearly the Earth must be stationary, and located right at the centre of the universe – and anybody who thought otherwise had better hope they didn't piss off the wrong Pope by saying so.

---

* The objection went thus: when we look at a star in the sky, we see it has a certain size, and, using fairly simple geometry, we can use that size and the distance from the Earth to figure out how big the star really is. Assuming the then-standard model of the universe was right, and stars were all pretty close to the Earth, then this calculation meant that most stars were comparable in size to our Sun, which would make sense.

If, however, Copernicus was right, and the Earth and planets revolved around the Sun in an unimaginably vast cosmos filled with mind-bogglingly distant stars, then that same calculation would imply that every star out there would be absurdly large – like, hundreds of times larger than our own Sun. This sounded ridiculous, even to Copernicans, who were forced to appeal to arguments like 'ah well, the Lord works in mysterious ways' to defend the idea.

It would be more than 200 years before the science necessary to resolve this paradox came about, and it was a fellow fallible genius that figured it out: George Airy. Around 1835, he published a paper describing the phenomenon now known as the 'Airy disc', which explains how, objectively, stars are just points of light in the sky, but when viewed through a curved lens – in, say, a telescope, or an eyeball – they get distorted, and appear to have a bigger size.

Now to be honest – and considering he was harbouring potentially blasphemous ideas in a time when doing so could find you on the wrong end of a death by burning alive – Galileo was in a pretty good position at this point. Copernicanism wasn't accepted by the mainstream, but it wasn't yet forbidden; thanks to Rheticus's hedging all those years ago, it was seen by most people as merely a purely theoretical mathematical model that somehow provided very accurate measurements and calculations of the heavenly bodies despite not having anything to do with the physical reality of the universe. But Galileo's observations, which he published in a small 1610 book called *Sidereus Nuncius*, or 'Starry Messenger', as well as the growing availability of good telescopes, slowly started to change that perception.

It wasn't enough for Galileo. Like any new scientific theory, his view that the Earth revolved around the Sun received pushback from the scholars of the age,* and that old sarcastic ego started to rear its head again.

'Oh, my dear Kepler, how I wish we could have a hearty laugh together,' he wrote to fellow astronomer and Copernican Johannes Kepler. 'Here at Padua is the principal professor of philosophy, whom I have repeatedly and urgently requested to look at the Moon and planets through my glass, which he pertinaciously refuses to do. Why are you not here? what shouts of laughter we should have at this glorious folly! and to hear the professor of philosophy at Pisa labouring before the grand duke with logical arguments, as if with magical incantations, to charm the new planets out of the sky.'

Frustrated at what he saw as a stubborn refusal from the establishment to see what was in front of them, he started accusing those who doubted him of '[believing] that God, when he created the heavens and the stars, had no thoughts beyond what they can themselves

---

* This is a crucial part of science, and not, as some tabloids may imply, a sign that one side or the other of the debate is plotting the downfall of civilization through the medium of, say, reusable straws.

conceive'. He started writing letters to other astronomers arguing that scientific observation should take precedence over Biblical lore – correspondences that made it into the hands of his, at this point, understandably disgruntled academic peers, who sent them on to the Inquisition. And then, in 1616, things got real.

In February 1616, prompted by Galileo's brazen and honestly kind of dickish promotion of heliocentrism, the Inquisition convened a commission adjudicate on Copernicanism. They concluded that the model was 'foolish and absurd in philosophy, and formally heretical since it explicitly contradicts in many places the sense of Holy Scripture'; Copernicus's treatise was banned, and Galileo was personally warned by the Church to shut his goddamn mouth.

'Aha,' you might be thinking at this point. 'This is the famous bit of Galileo's life – you know, the bit where the mean old Pope comes along and forces him to recant the scientifically true position of heliocentrism before condemning him to house arrest for the remainder of his life.' But almost unbelievably, Galileo was still being extended the benefit of the doubt: in fact, he had actually met the Pope just a few weeks after the judgement came down, and was assured that he would not be prosecuted for his behaviour.

So, for a while, Galileo behaved himself. But by 1623, a new Pope had been elected – and Galileo decided to push his luck again.

To be fair, you can probably see why he thought he was safe. The new Pope, Urban VIII, had previously been known as Maffeo Barberini, and he had actually known Galileo for more than a decade by the time he ascended to the papacy. They had originally met at a dinner in 1611, where Barberini had immediately become a fan of Galileo's unbeatably sardonic debate style. He had even allowed Galileo to publish a new book, *The Assayer*, and had reportedly enjoyed it so much that he had been heard roaring with laughter at Galileo's snarky lambasting of the Jesuit scholars who rejected mathematics and science in favour of theistic philosophy. Perhaps, then, he could be convinced on Copernicanism?

At first, things looked good for Galileo: he met with the Pope to discuss his theories, and succeeded in coming away with formal permission to write about the theory once again. He would have to compromise, though: he was not to write a treatise, but a dialogue – a book in the form of two characters, each proposing one view of the universe, and debating the strengths and weaknesses of each. The work would be called the *Dialogue Concerning the Two Chief World Systems*.

What could go wrong? Let's see if you can spot Galileo's super subtle slant on the topic. First, the character representing the Copernican view of the universe was named Salviati, while the character who represented the Aristotelian view – that is, the view espoused by the Church – was named Simplicio, supposedly named after Simplicius of Cilicia, an Aristotelian writer from the sixth century, but I mean ... come on. It meant the same then as it does now: the Simpleton. Salviati was written as intelligent; Simplicio was foolish and stubborn, and worst of all, some of Simplicio's arguments came almost verbatim from the mouth of Pope Urban VIII himself.

Galileo's devastating one-liners – his trump card that had served him so well professionally, even when his mathematics was technically lacking – had backfired. In his eagerness to publish his Copernican ideas, he had forgotten a cardinal rule: if you're living in a Catholic country in the sixteenth century, *don't call the Pope an idiot*.

Now nearing seventy, Galileo was hauled in front of the Inquisition once again, and this time there was no 'private word' – he was to defend himself in person at the Vatican. He was accused of violating the 1616 order against teaching Copernicanism as fact, and threatened with torture; in response, he appealed on technicalities. *Technically*, he pointed out, a quirk of bureaucracy had meant that no official injunction had been issued back in 1616, and even if it had, he said, he hadn't taught it as fact – he had presented both sides of the argument, as instructed, and it could hardly be *his* fault if one perspective came across as more convincing than the other.

In the end, it was his daughter Virginia – now going by the name of Sister Maria Celeste – who broke him. They had remained in contact throughout her life, and held a deep respect for each other; she urged him to submit to the Inquisition, and he finally relented. He agreed that he had been too arrogant and promoted Copernicanism too forcefully, offering to write an extra hundred pages of the *Dialogue* to rectify the situation. Throwing himself upon the mercy of the Inquisition, which has not often proved to be a good idea, historically speaking, he knelt before them and proclaimed that 'with a sincere heart and unfeigned faith I abjure, curse, and detest the said errors and heresies, and generally every other error and heresy contrary to the ... Holy Church, and I swear that I will nevermore in future say or assert anything ... which may give rise to a similar suspicion of me.'

The Inquisition sentenced Galileo to 'imprisonment at the pleasure of the Holy Congregation', and the Pope demanded he publicly renounce his views. He lived the rest of his days – eight years in total – under house arrest, watching his ideas take over the world from a hilltop villa in Florence, and honestly – that sounds pretty bearable as prisons go.

You might expect that those last eight years would be spent ruing his lack of tact, but you'll be happy to hear that Galileo absolutely did not learn his lesson. Not only did he continue conducting scientific experiments and astronomical observations, he wrote another book summarising his discoveries: the *Dialogues Concerning Two New Sciences*. Salviati and Simplicio were back, baby, and this time they were talking about engineering.

The Inquisition had a long and unforgiving memory, though, and Galileo was still not allowed to publish any new work. Just like today, though, rules in the sixteenth century didn't exactly apply to celebrities in the same way as they do to the rest of us, and he was able to get his manuscript over to a printing company in Holland, which had been Pope-free for about fifty years by this point.

Within a year, the book had reached Rome, where it sold out immediately. And that's the story of how Galileo eventually, and extremely circuitously, got the better of the Inquisition.

5.

The Entirely Unbelievable Life of Tycho Brahe

If you ask your physics or maths teacher who Tycho Brahe was, they will probably say something like this: 'Well, Jimmy [your name is Jimmy in this hypothetical], Tycho Brahe was a sixteenth-century Danish nobleman and astronomer who discovered supernovas and created the most accurate star charts ever known at that point. Without him, Jimmy, we wouldn't have had Kepler's laws of planetary motion, or Newton's theory of universal gravitation, or perhaps even a scientific revolution at all.'

This is wrong. Well, it's not wrong at all, actually, but it's definitely not the whole picture. A better way to describe Tycho Brahe would be something like 'what you get when every atom of chaos energy in the universe converges on to one super-rich Danish guy', or if you want to get hyper-specific, 'a feudal lord who lived with a drunk moose and a psychic dwarf. And he had a brass nose.'

Now, with some of history's more eccentric figures, you can tell from a young age that things are going to end up a bit weird for them. This isn't the case with Tycho Brahe – not because his childhood was normal or even normal-adjacent in any way, to be clear, but because the madder aspects of his life were set in motion long before he was born. See, there's this thing that noble people in the olden days used to like to do, where they would decide who to marry and have kids with based not on the things we think are important today, like 'makes me laugh' or 'has an arse that won't quit', but on more prosaic attributes, like 'owns Sussex' or 'uncle is a duke'. The problem with this approach, though – other than the fact that you're likely to be cast as an antagonist in a period rom-com a few hundred years after your death – is that you're pretty quickly going to dry up the gene pool, and after that it's nothing but Habsburgs.

Tycho was born in 1546, basically as the last possible step before incest: he was heir to no fewer than fourteen noble Scandinavian families, a fact that made him less *a* Danish nobleman and more *the* Danish nobleman. It also made him ludicrously rich, and not just in that 'I technically own a whole bunch of peasants and the entire county they live in' way that people could be back in those days; at one point it was estimated that he was worth fully one per cent of the GDP of his entire country, a milestone not even the richest people on the planet have yet managed to repeat.

He was also born a twin – though his brother died either just before or just after birth, nobody's quite sure. He was the second child, but first son, of Otte Brahe and Beate Bille, a pair who obviously liked each other very much since they went on to have ten more children after Tycho.

This may be why they didn't kick up too much of a fuss when, aged just two, Tycho was kidnapped in broad daylight by Otte's brother Jørgen. Instead of the 'No! Please! We'll give you anything!' that you would expect from parents learning their first-born son had been stolen from them, Otte and Beate reacted more like 'Ah, ya got us, OK, fair enough' and just kinda accepted the child-theft; it turned out that Otte, knowing his wife was expecting twins, had promised one of the boys to his childless brother and sister-in-law, and so as far as Jørgen was concerned, this so-called 'criminal child abduction' was just him taking what was rightfully his.

Still, other than a gazillion siblings running around, Jørgen and his wife Inger provided their adopted nephew-son with everything he would have got at home – Tycho would later write that Jørgen 'raised me and generously provided for me during his life until my eighteenth year; he always treated me as his own son and made me his heir.' Tycho was given a first-class education, which he apparently must have excelled at since he enrolled at the University of Copenhagen aged just twelve, and at fifteen he was sent on a tour around Europe in the hopes that this would inspire him to become some kind of civil servant like basically every one of his forebears. Sadly for Jørgen, though, his dreams of being the uncle-father of a governmental advisor or something would never be realised, for two reasons: firstly, because there was a total solar eclipse on 21 August 1560, and secondly, because Jørgen went for drinks with the King in June 1565, drunkenly dived into a canal to rescue the equally wankered King from drowning, and ended up dying of pneumonia a couple of weeks later.

The solar eclipse, though, realigned Tycho's life completely. He had always been more interested in science than Jørgen would have liked, but the day he witnessed the Moon passing in front of the Sun and plunging Europe and Africa into darkness for nearly four minutes, he was fully converted. I mean, come on: even today, it's pretty awe-inspiring, and *we* know what's going on. Back in Tycho's day, the sight of the Sun disappearing from the sky was seen as nothing short of apocalyptic; birds 'fell to the ground', and fighting broke out in churches as people climbed over each other to confess their sins, which must have been fairly embarrassing for all involved about five minutes later.

For Tycho, though, the solar eclipse wasn't an omen, but a promise – it heralded humanity's growing mastery over nature. What captured his imagination the most was not the disappearance of the Sun – having been a disappeared son himself a mere decade ago, this can't have been that impressive to him – but the fact that it had been predicted in advance.

Even more tantalisingly, it had been predicted *badly*, arriving a day out from the expected date. Tycho, thirteen years old and rich as Croesus, took a look at his almanac, looked at the obscured Sun, and made the decision so many other second-year undergrads have come to throughout history: that making a breakthrough significant enough to entirely revolutionise the way science is thought of as a concept ... would be easy, probably.

From that point on, Tycho devoted his life to science – or at least, the closest that sixteenth-century Europe came to science, which is to say, he studied a whole bunch of stuff like astrology and alchemy and how to cure a mild headache using tinctures of lead and mercury. He enrolled at the University of Rostock, in Germany, and in the tradition of students everywhere, got drunk a lot – and if you know just one fact about Tycho Brahe then you probably know what's coming up: the nose thing.

In December 1566, aged twenty, Tycho went to a party. It was at the home of one of his lecturers, so you might expect him to have at least *tried* to behave himself, and maybe everything would have been fine if his cousin Manderup Parsberg hadn't been there too. Surrounded by booze and festivity, though, the pair soon got into a heated argument over which of them was the superior mathematician.

Now, of course, getting into arguments with your cousin about who's better at maths is to be expected at a party thrown by a theology professor, but this was no standard maths-based grudge match. For one thing, it lasted nearly three weeks: the party was held on the 10th of December, and the argument wasn't resolved until the 29th. But mostly, it's gone down in history because of the *way* it was resolved, which you might assume was by some kind of exam-based maths-off, but in fact took the form of rapiers at dusk.

But as anybody who's tried it (or, you know, thought about it for more than a few seconds) can tell you, sword fighting in the dark is a bad idea, which is why Tycho left the duel with a large cut across his forehead and no nose. Luckily, being obscenely wealthy and the heir to several of the most elite families in the kingdom tends to incentivise folk to keep you alive, and Tycho received the best care possible, which in this case means that he went down in history as 'the scientist with the brass nose'.

Within a few years, Tycho had graduated from Rostock and returned home, where his father – his real father, that is, not his uncle-father – was dying. Now a die-hard astronomer, he started putting his enormous fortune to use, singlehandedly building an astronomical observatory and alchemical lab where he could continue his studies, or at least singlehandedly paying for other people to build it for him, and further bucked the family tradition soon afterwards by falling in love with a woman named Kirsten Barbara Jørgensdatter.

Kirsten was the daughter of a minister, and therefore a commoner. While this would have been good news for Tycho from a genetic point of view, it was bad news if he wanted to actually marry Kirsten, because Danish law at the time forbade commoners from becoming nobles. This meant that if the two had tied the knot, instead of her becoming posh, as you might expect, he would become common, thereby losing a bunch of money, land and privilege. While he loved Kirsten, it seems he didn't love her 'one per cent of the GDP of Denmark' worth, and the pair were left with three options: split up and go their separate ways, for ever wondering what might have been; die tragically in a Romeo-and-Juliet-style elopement gone wrong; or say 'fuck the haters' and shack up together regardless of what society said.

Of course, one of the best things about owning just *all* the money is that it becomes a lot easier to go that third route, and so that's what they did: they entered into a 'morganatic' or 'left-handed' marriage, which basically meant that they lived together as if they were husband and wife for so long that everyone had to accept they may as well count as married. Following in his father-father's prolific footsteps, Tycho had eight children with Kirsten, none of whom were initially allowed to take his name or inherit his land and titles.

But then, for some reason, the King of Denmark decided that a hot-headed astronomer with a brass nose and a handful of illegitimate children would be the perfect representative for the Crown abroad, and Tycho was sent on another tour of Europe.

Ostensibly, though, he was travelling for himself. He went first to Germany, then Switzerland and Italy, visiting the continent's most famous observatories; he had recently discovered the existence of a supernova in the constellation Cassiopeia, and his reputation as an astronomer was now known internationally. That's not to say everybody supported him – it had been taught since ancient times that the heavens were eternally fixed, so his announcement that there was a new star in the sky was verging on scientific heresy. But Tycho gave little thought to his detractors, addressing them in the preface of his treatise on the *stella nova* as 'O crassa ingenia', which is Latin for 'hey, thickwits'. (Seriously. Look it up.)

This sightseeing and Danish charm offensive trip lasted only about a year, at which point Tycho returned home and started making plans to move to Switzerland. 'I am displeased with society here,' he told his friend Johannes, complaining about the 'customary forms and the whole rubbish', which is fair when you remember how he was banned from marrying his wife or letting his kids inherit anything from him. But the King got wind of his intentions and bribed him to stay in Denmark instead – and when a sixteenth-century king wants to bribe you, he can really make it worth your while.

Tycho was given an entire island, Hven, which he basically set about ruling as a sort of mini king. He exacted taxes from the families who lived there, which they weren't too happy about, but was honestly the least of their problems, since they had just been unilaterally demoted to serfs and Tycho was not above forcing them to perform hard labour for him for free, a practice we know today as 'slavery'. The Hvenians took Tycho to court over his treatment of them, which really proves that they had no idea what serfdom was all about because of *course* the King sided with Tycho, and all they had done was set a legal precedent by which this guy with a precariously-glued-on nose and a huge scar across his face gets to order them to do whatever he wants.

Before too long, Tycho's observatory-cum-palace – named Uraniborg, after the muse of astronomy Urania, not the Delta quadrant antagonists The Borg – was finished, complete with underground observatory, in-house printing press, and working dungeon in which to throw any peasants who decided to complain about being forced to build some fancy-pants castle for zero wages.

And that's when things started to get weird.

When Tycho wasn't studying the heavens or compelling local farmers to be his slaves, he would spend his time partying. And trust me when I say: *nobody* parties quite like a drunk-arse sixteenth-century nobleman with their own autocratic island, enough money to buy Wales, and the constant aura of pure chaos that followed Tycho his entire life. And for proof, look no further than his poor pet moose.

Now, if you're not familiar with mooses* – and not many people are, since their natural habitat is confined to a single strip in the most Northern latitudes of the world – let me explain something: they're *huge*. A fully grown male can easily reach three metres from floor to antlers, which, by the way, are nearly two metres wide in their own right. They can weigh well over half a tonne, they regularly kick wolves and bears to death, and they can run at 35 mph if they want to.

If that's hard to gauge in real-world terms, let me put it this way: a moose would not be able to stand up in your bedroom, and it *would* be able to outrun your car in an urban zone, then turn around and smash it to pieces.

All of which is to say: you probably wouldn't want to get one drunk. Unless you're Tycho Brahe, that is, who apparently considered the optimum state for a housetrained moose was 'absolutely slaughtered at all times'.

---

* Or to use the scientific term, *meese*.

It says a lot about Tycho's eccentricities that the weird part of him owning a pet moose wasn't *that he owned a pet moose*, but here we are: this moose trotted alongside Tycho's carriage when he went on trips, slept at the foot of Tycho's bed, and drank Tycho's beer and mead until it was staggering through Uraniborg grunting and bellowing the moose equivalent of 'Show Me the Way to Go Home'.

And Tycho, who, to be fair, was something of a drunken moose himself, thought this was *absolutely hilarious*, and instead of sending his beloved pet to AA or, you know, just taking a good hard look at his life, he would lend it out to his friends for entertainment at parties like some kind of cervine pimp.

One of those friends was the Landgrave of Hesse-Kassel, Wilhelm IV. He had written to Tycho in the spring of 1591 to ask whether the animal-lover had ever heard of a 'rix' – a creature that looked, he had been told, quite like a small deer, but both stronger and nimbler, and with shorter antlers. Tycho replied that such an animal didn't exist as far as he knew, but had the landgrave ever seen a drunk moose?

Despite a moose being pretty much the opposite of 'like a deer but smaller and less antlery', Wilhelm was immediately interested, and Tycho promised to send it over. Unfortunately, though – or at least, arguably, unfortunately – the landgrave never got to see his home demolished by a ten-foot-tall blackout-drunk housemoose, because just a couple of days later, Tycho and Moose were invited to a party closer to home.

As so often is the case when two raging alcoholics are put in a room with unlimited booze, catastrophe ensued. At some point in the night, the moose, following the ancient instinct of drunk partygoers everywhere, decided to wander upstairs and explore – an activity that anybody can tell you is difficult enough on two legs, let alone four. The moose lost its footing, and fell, crashing down the steps and dying a few days later.

The moose was far from the only animal that lived with Tycho at Uraniborg, but it was definitely his favourite, and Tycho was heartbroken at this loss – not just of a pet, but of a loyal drinking buddy.

Still, at least he had Jeppe. Jeppe was not a pet, although you wouldn't know that if I hadn't specified it: he sat at Tycho's feet throughout mealtimes, occasionally being fed scraps from his master's plate as he chattered away to himself in a language only he understood. Other times, he would dance for the partygoers, and perform tricks on command. He truly was Tycho's favourite and most loyal companion.

Look, there's no real way to sugar-coat this: Jeppe was a psychic dwarf that Tycho kept as a slave. It's one of those obviously terrible things that rich people did in the olden days, but if you were anybody of any renown in the Old World, chances were you'd have a dwarf. It would have been quite a brutal trade-off for the dwarfs themselves: they'd get to live in a palace, eat the best food, wear fine clothes, and in some cases would even be allowed to mock and poke fun at their lords – no small boast in a feudal society. But in return, they would often be treated less like people and more like toys or pets, being forced to dress in stupid costumes or to perform demeaning stunts.

This dichotomy was pretty much as obvious as it could possibly be with Jeppe. Yes, he was treated essentially like a particularly talented dog for the most part by his less-sympathetic-by-the-paragraph master – but equally, he held power over Tycho that probably nobody else in the world could boast.

That's because Tycho was absolutely convinced Jeppe was clairvoyant. And apparently for good reason: ask Tycho to prove his minion's psychic abilities, and he would point to the time he was awaiting the return of two of his research assistants from Copenhagen, only for Jeppe to suddenly stand up in the middle of dinner and announce: 'See how your people are laving themselves in the sea.'

Immediately, though presumably after finding a dictionary and looking up the word 'laving', Tycho sent a servant to the roof of the castle to look out to sea and figure out what Jeppe was talking about – and would you believe it? There was indeed a boat out there, upside down in the water next to two soaked, though thankfully alive, assistants.

And Jeppe also served as a sort of morbid eight-ball, being taken around to Tycho's friends when they fell ill and predicting whether they would survive or not. Apparently, he was never once wrong, although this is less impressive when you remember he was making these determinations in a time when a mild flu or a toothache could kill you, and so even if he got a prediction wrong he could probably, ahem, rectify it without raising too many suspicions.

But let it not be said that Tycho didn't put his slave-built party palace to academic use. In 1588, he published the second of his two-volume work, *Astronomiae Instauratae Progymnasmata*, or 'Introduction to the New Astronomy'. It was, as you'd expect, finished about fifteen years *before* volume one, and it contained within it a full explanation of Tycho's model of the solar system, a theory which was highly complex, slightly heretical, and completely wrong in just about every aspect. His astronomy palace now having served its purpose, he packed up his family, his instruments, and his dwarf, and left Denmark for the Holy Roman Empire, a trip that had nothing at all to do with the fact that the new king didn't like him and his peasants were revolting at his palace doors.

Eventually, the Brahe family settled in Prague, where Tycho found what would become his most famous legacy: a young German physicist by the name of Johannes Kepler.

On paper, the two couldn't be more different: Kepler came from poverty – he was the son of a witch* and a mercenary – and was

---

* Katharina Kepler was, along with fourteen other women, accused of witchcraft in 1615 in Stuttgart – an accusation she and her son vehemently denied, for obvious reasons. Since due process wasn't really a thing for women back then, and seeing as how eight of those accused had already been executed, Johannes gave up trying to lawyer her out of it and took her to Linz, a city that was almost 500 miles away and therefore as good as vanishing off the face of the Earth in 1600s money. For some reason she decided to go back home a few years later, where she was promptly locked up and threatened with torture until Johannes saved her again.

self-deprecating to an extent that made you feel bad for the guy. Not the kind of man who would keep a little person as a psychic pet, is the point. The pair disagreed scientifically, too, with Kepler favouring the (correct) Copernican model over Tycho's own model; they even disagreed on how to *do* science at a fundamental level, which for Tycho was to observe and record and for Kepler involved the construction of elegant theories that explained the universe on a holistic level. Unsurprisingly, their partnership was a fraught one, with full-blown arguments being the norm for months until eventually Kepler stormed out in a rage.

But once egos had been swallowed and apologies had been made, the pair found that, actually, they had a lot in, well, not in common, but at least in complement. Tycho's ultra-fastidious record keeping allowed Kepler to keep his theories in check, eventually inspiring his Laws of Planetary Motion. Kepler, on the other hand, was just about the only research assistant who could keep up with Tycho's demand for accurate data – and when it came to Mars, the most confusing planet in the skies, he out-measured even the old master.

So impressed was Tycho with his serendipitous apprentice that he promised to recommend Kepler for the position of Imperial Mathematician, which was presumably like being a poet laureate but with equations. Kepler's life would have been the ultimate physics rags-to-riches tale … had Tycho not, in October 1601, gone to another party.

In the end, Tycho Brahe died the way he had lived: drunk, stupid and probably hungover. There's been so many conspiracy theories and stories about his death throughout the years – maybe he was murdered for having an affair with the Queen of Denmark, or maybe Kepler poisoned him for being too much of a jerk at the office; maybe he killed himself, accidentally, through improper lab safety protocols in the alchemy dungeon.

We'll probably never know for sure, but the generally accepted theory is that he pissed himself to death – or, rather, that he *didn't* piss himself to death. According to Kepler, Tycho refused to go to the toilet throughout that final shindig, preferring instead to wait until he got home. But once he reached his own privy, he found he basically couldn't go at all, and those drips he did manage to squeeze out caused him extreme pain. He spent the next night deliriously begging Kepler to embrace the Tychonic model of the solar system instead of Copernicus's heliocentric version, wrote himself an epitaph – 'he lived like a sage and died like a fool', which is arguably overselling how he had lived – and died, making the moose falling down the stairs look graceful by comparison.

Never one to do things like the common folk, Tycho would be buried three times over. The first time was, well, you know, the normal one, but it didn't take: in 1901, he was dug up again and given an autopsy.

Now, three centuries after death is normally considered just a smidge too late for a post-mortem, but they did it anyway, and apparently only so they could rub it in the face of Tycho's equally long-dead doctor, who had diagnosed the cause of death as kidney stones back in 1601. Since no stones were found at the autopsy, the doctors decided he probably died from something to do with his prostate.

Tycho then spent another century or so mouldering peacefully in the ground, until in 2010 he was exhumed once again. This time, the reasoning was more *CSI* than *House, MD* – the researchers wanted to disprove all those 'murdered by a jilted mathematician' rumours surrounding his death. They took tooth, bone, hair and clothing samples and analysed them for mercury, which was the Tycho-truthers' poison of choice, as well as various other toxic substances, and they discovered two things: first, that there was no way he was poisoned by anything, and secondly, that he was basically gold plated.

Literally – his hair contained clear traces of the stuff, showing a hundred times the normal levels of gold in his body. The researchers concluded that he must have been exposed to gold pretty much every day of his life in order to have built up the amount they found in his body. He was probably even eating and drinking it, by way of gold dining plates and gold leaf in his wine, which, let's face it, is absolutely the most decadent thing either of us have heard of – and probably, in his underground Uraniborg labs and lairs, he was mixing up and consuming gold-based potions and tinctures as well. But while heavy-metal poisoning was, let's say, a not uncommon cause of death for many alchemists at the time, it wasn't what took old Tycho's life. With his death re-mystified, and his reputation firmly established as some sort of proto-C-3PO, the team buried Tycho for the third and, so far, final time.

And thus, for now, ends the story of Tycho Brahe, the man too crazy to be believed – and too rich to be stopped.

6.

When René Descartes Got Baked

René Descartes, like Pythagoras before him and Einstein after, occupies that special place in our collective consciousness where his work has become ... well, essentially a short-hand for genius-level intellect.

Think about it – in any cartoon or sitcom where one character is (or, through logically-spurious means, suddenly becomes) a brainiac, there are three things they're narratively bound to say: 'the square of the hypotenuse is equal to the sum of the squares of the other two sides' – that's Pythagoras; '$E = mc^2$' – thank you, Einstein; and finally, 'cogito ergo sum'. And *that* is Descartes.

Specifically, it's *old* Descartes – Descartes after he had figured his shit out. But while his later writings undeniably played a huge and important role in setting up how we approach the world today – he's actually one of the main figures who brought us the concept of the scientific method – Descartes's early years leaned a little more on the silly and gullible than the master of scepticism he's come to be known as.

Descartes was born in 1596, which places him firmly in that period where science and philosophy and magic were all pretty much the same thing. He's probably best known as a philosopher these days, but that's likely because a lot of his developments in mathematics have become so incredibly fundamental that we kind of forget they had to be invented by anybody at all. And I know I'm saying that with ten years of mathematical training behind me and a PhD on the shelf, but even if you haven't set foot in a maths class since school, you'll be familiar with something that Descartes invented, because he was the guy who came up with graphs. That's actually why the points in a graph are given by *Cartesian coordinates* – it's from the Latin form of his name, Renatus Cartesius.

And while maths, despite what everyone keeps telling me, *can* be sexy, 'cogito ergo sum' really does have a nice ring to it, doesn't it? 'I think, therefore I am.' It doesn't sound like a huge philosophical leap – in fact, it kind of sounds like tautological nonsense – but it's actually one of the most important conclusions ever reached in Western thought.

See, before Descartes, philosophy didn't exactly have the sort of wishy-washy, pie-in-the-sky reputation it enjoys today. The dominant school of thought was Scholasticism, which was basically like debate club mixed with year nine science. Sounds fair enough, but in practice – and especially when combined with the strong religious atmosphere and general lack of science up till that point – it was basically a long period of everybody riffing on Plato and Aristotle and trying to make their Ancient Greek teachings match up with the Bible. This was, needless to say, not always easy, and led to rather a lot of navel gazing over questions like 'Do demons get jealous?' and 'Do angels take up physical space?'

Descartes's approach was radically different. He didn't see the point in answering questions like how many angels can dance on the head of a pin until he'd been properly convinced of the existence of angels. And dancing. And pins.

Now, of course, this is the point when non-philosophers throw up their hands in despair and say something along the lines of 'Of *course* pins exist, you idiot, I have some upstairs keeping my posters up! Jesus, René, are we really paying a fortune in university fees just so you can sit around and doubt the existence of stationery?'

But to that, Descartes would reply: are you sure? I mean, we've all had dreams before that are so convincing that we wake up thinking we really *did* adopt a baby elephant after our teeth all fell out. How do I know I'm not dreaming now? How do I know this isn't a *The Matrix*-type situation, and what you think are pins are just a trick being played on us by Agent Smith?

In fact, when you get right down to it, Descartes would say, how can we be sure *anything* exists? I might not even exist! I might be a brain in a vat, being cleverly stimulated in such a way as to induce a vast hallucination! And yes, sure, I agree that sounds *unlikely*, but it's not *impossible* – the point is, we simply can't *know*.

The only thing I can be sure of, Descartes would continue – despite everyone by this point rolling their eyes and muttering things like 'see what you started, Bill' – is that *I* exist. And I can be sure of that, because I'm thinking these thoughts about what exists. I may just be a brain in a vat, being fed lies about the reality that surrounds me, but 'I', 'me', my sense of self and consciousness – that definitely exists. To summarise: I think – therefore I am.

It was a hell of a breakthrough – he'd basically Jenga'd the entire prevailing worldview into obsolescence. And it's the kind of idea that could really only have come from someone like Descartes: a weirdo celebrity heretic pseudo-refugee who had a weakness for cross-eyed women, weed and conspiracy theories.

Descartes was, as his name suggests, French by birth, hailing from a small town vaguely west of the centre of the country. If you look it up on a map, you'll see it's actually called Descartes, but it's not some uncanny coincidence – the town was renamed in 1967 after its most famous resident.

Which is kind of odd, because it's not like Descartes spent all that much time there. He went to school in La Flèche, more than 100km away, where even at the tender age of ten he was displaying the sort of behaviour that would make him perfectly suited to a life of philosophy, sleeping in until lunch every day and only attending lectures when he felt like it. This can't have made him all that popular with the other kids, who were all expected to get up before 5am, but that's why you choose a school whose rector is a close family friend, I suppose, and, in any case, by the time the young René turned up they were probably all too tired to do much about it.

After finishing high school, he spent a couple of years at uni studying law, as per his father's wishes – his dad came from a less well-to-do branch of the Descartes family tree, and probably would have wanted Descartes to keep up appearances for the sake of holding on to posh perks like not paying taxes. It must have pained him, therefore, when after graduating with a *Licence* in both church and civil law, Descartes immediately gave it all up and went on an extended gap year.

'As soon as my age permitted me to pass from under the control of my instructors, I entirely abandoned the study of letters, and resolved no longer to seek any other science than the knowledge of myself, or of the great book of the world,' he would later write, like some kind of nineteen-year-old *Eat Pray Love* devotee.

'I spent the remainder of my youth in travelling, in visiting courts and armies, in holding intercourse with men of different dispositions and ranks, [and] in collecting varied experience,' he continued, in his philosophical treatise-slash-autobiography *Discourse on the Method of Rightly Conducting One's Reason and of Seeking Truth in the Sciences*, which for obvious time-saving reasons is usually referred to as *Discourse on the Method*. And like so many philosophy students throughout history, there was one place he found in his travels that caught Descartes's heart and imagination more than anywhere else: Amsterdam.

Now, it is of course true that places can change a lot over the course of 400 years – at this point in history, France was being ruled by a nine-year-old autocrat and his mum, Germany didn't exist, and England was a few years short of becoming a Republic. So you might think, sure, *these days* Amsterdam has a bit of a reputation, but back in Descartes's time, it was probably a hub of quiet intellectualism and sombre, clean living.

Nope! Dynasties may rise and fall, empires spread and eventually fracture, but apparently, Amsterdam has always been Amsterdam. Descartes spent his first few years in the city living his absolute best life, studying engineering and maths under the direction of Simon Stevin – another guy you've never heard of who made a mathematical breakthrough you almost certainly use every single day of your life, since he invented the decimal point – and dressing like an emo and throwing himself into music. He joined the Dutch army for a bit, despite being by all accounts a tiny weedy bobble-headed French guy, and, yes, he almost certainly smoked a bunch of pot along the way.

And then, one November night in 1619, while on tour in Bavaria, Descartes had a Revelation. And he had it, according to his near-contemporary biographer Adrien Baillet, inside an oven.

'He found himself in a place so remote from Communication, and so little frequented by people, whose Conversation might afford him any Diversion, that he even procured himself such a privacy, as the condition of his Ambulatory Life could permit him,' Baillet writes.

'Not … having by good luck any anxieties, nor passions, within, that were capable of disturbing him, he staid withal all the Day long in his stove, where he had leisure enough to entertain himself with his thoughts,' he continues, as if that's a normal thing to write and not an account of someone being so introverted that they secluded themselves miles away from anyone who knew them and then crawled into an oven for the day.

Modern biographers have suggested a few interpretations of what this oven might have been, and I'm sorry to report that, of course, it's not as ridiculous as it first seems: in the seventeenth century, before we'd tamed electricity and gas mains and whatnot, a 'stove' or 'oven' was more like your modern-day airing cupboard than an Aga. Just bigger. And fancier. And all your towels are on fire. Look, the analogy isn't perfect, but the point is that when Descartes said, in *Discourse on the Method*, that he had 'spent all day entertaining his thoughts in an oven', he wasn't being completely absurd – just, you know, kind of weird.

Depending on where you fall on the scale between 'Descartes was a stoner lol' and 'Descartes was a paragon of virtue, 10/10 no notes awesome dude', what happened next was either the result of too much weed, too much oven, or too much being a fricking genius destined to reform all of Western philosophy. Either way, he had a pretty rough night, full of strange dreams and disturbing hallucinations* that even the loyal Baillet thought might be a sign he was going a little bonkers.

'He acquaints us, That on the Tenth of *November* 1619, laying himself down *Brim-full of Enthusiasm*, and … *having found that day the Foundations of the wonderful Science*, he had Three dreams one presently after another; yet so extraordinary, as to make him fancy that they were sent him from above,' writes Baillet, just in case you were wondering where on that scale Descartes would put himself. In fact, so sure was he of the divine nature of his dreams that, Baillet said, 'a Man would have been apt to have believed that he had been a little Crack-brain'd, or that he might have drank a Cup too much that Evening before he went to Bed.

'It was indeed, St. *Martin*'s Eve, and People used to make Merry that Night in the place where he was … but he assures us, that he had been very Sober all that Day, and that Evening too and that he had not touched a drop of Wine for Three Weeks together.'

Sure, René. Though honestly, the content of the dreams aren't as noteworthy as the conclusions he drew from them – unless you think

---

* Some modern scientists have suggested that Descartes's night in the oven may in fact be the earliest recorded experience of Exploding Head Syndrome, a sleep disorder you may well have had yourself once or twice. Despite the gnarly name, it doesn't actually involve your head exploding – that would certainly have made Descartes's future work more impressive – but it does cause you to hear loud bangs and crashes that aren't really there, and sometimes see flashes of light as well, both of which Descartes recorded experiencing that night.

'walking through a storm to collect a melon from a guy' is super weird, I guess. And goodness knows how he got from cantaloupe to conceptualism, but these three dreams are said to have given him the inspiration first for analytic geometry – that is, his maths stuff – and then the realisation that he could apply the same kind of logical rigour to philosophy.

And I don't want to minimise what Descartes achieved after this melon-based enlightenment – it takes guts to stand up in a world governed by strict ritual and belief and announce that not only is everyone around you an idiot, but also they probably don't even exist, so there. But have you ever heard that saying about not being so openminded that your brain falls out?

Well, 1619 was also the year that Descartes, writing under the pseudonym 'Polybius Cosmopolitanus' – Polybius being an ancient Greek historian, and Cosmopolitanus being Latin for 'citizen of the world' – released the *Mathematical Thesaurus of Polybius Cosmopolitanus*. It kind of sounds like a Terry Gilliam movie, but it was actually a proposal for a way to reform mathematics as a whole.

It doesn't matter that you've never heard of it. It's not as famous as the *Discourse*; in fact, it may not have ever even been completed. The important bit wasn't what was contained inside the book, but who it was dedicated to: to 'learned men throughout the world, and especially to the F.R.C. very famous in G[ermany].'

And who was this mysterious F.R.C? Descartes was specifically referencing the *Frères de la Rose Croix*. In English, they were known as the Brothers of the Rosy Cross – and, today, they're called the Rosicrucians.

So, you may have heard of the Rosicrucians, but it's more likely you haven't. Today, the term actually refers to two separate organisations, both of which claim to be the 'real' Rosicrucians and both of which denounce the other group as being a bunch of weirdos. They're equally wrong on the first point, and equally right on the second: there's no Rosicrucian group around today that is directly linked to the original group that Descartes was a fan of, and every iteration of the organisation is and always has been fucking bananas.

But people in search of a new outlook on the universe often don't get to choose which batshit philosophy the world throws at them first, and Descartes had the peculiar fortune of going through his minor mental breakdown in early seventeenth-century Germany.

Between 1614 and 1616, three 'manifestos' were published in Germany. They were anonymous, recounting the tale of one Christian Rosenkreuz, a man who was born in 1378, travelled across the world, studied under Sufi mystics in the Middle East, came back to Europe to spread the knowledge he had gained in his travels, was rejected by Western scientists and philosophers, and so founded the Rosicrucian Order, a grand name for what was apparently a group of about eight nerdy virgins. All of this, the manifestos said, he accomplished by the age of about twenty-nine, after which he presumably just sat on his thumbs for a long old while since the next big thing he's said to have done was die aged 106.

Now, some people have posited that everything you just read is false – a kind of early modern conspiracy theory. And yes, 'Christian Rose-Cross', as the name translates from German, is rather on the nose for the founder of a Christian sect, and, *yes*, it's a bit farfetched for anybody to have lived for more than a century in the 1400s, *and*, *yes*, OK, so the last manifesto was almost certainly actually written by a German theologian named Johann Valentin Andreae, who was attempting to take the piss out of the whole thing and publicly renounced it when he realised people were taking him seriously – but that's the thing: people *did take it seriously*. And one of the people who took it seriously seems to have been Descartes.

'There is a single active power in things: love, charity, harmony,' mused the philosopher most famous for radical doubt of everything that couldn't be proved via logic alone. Not in any published work – these were the thoughts of Descartes the early-twenties guy just trying to figure his shit out, found years later in the journal he kept throughout his life.

Another: 'The wind signifies spirit; movement with the passage of time signifies life; light signifies knowledge; heat signifies love; and instantaneous activity signifies creation. Every corporeal form acts through harmony. There are more wet things than dry things, and more cold things than hot, because if this were not so, the active elements would have won the battle too quickly and the world would not have lasted long.'

If that sounds, you know, completely ridiculous to you, that's probably because we live in a post-Descartes world, and he didn't. All this poor oven-baked idiot had at his disposal were a dream about melons, a steadfast conviction that he had been personally chosen by God to reform the entirety of Western thought up until that point, and some rumours about a weird sect of rosy German virgins who were devoted to doing just that.

You may have already guessed the next bit of the story: Descartes joins the Rosicrucians and embarks on some insane rituals and philosophies that we've never heard of today because it doesn't fit in with our modern ideas of 'genius', right?

It's actually way more stupid than that. In a series of events that, once again, really feels like it was ripped straight out of some cult comedy movie, Descartes tried to join the Rosicrucians, but kept running into the problem of them not, in fact, existing. So he couldn't join the group, but what he could and did do was accidentally make everyone *think* he had joined, thus entirely screwing over his reputation as someone to take seriously.

Of course, in the grand scheme of things, this didn't matter much, because to a lot of people he was dangerous enough even without all the conspiracy stuff: his insistence that truth was something for humans, not God, to judge, and the idea that authority should or even could be questioned, made him an enemy of most established Churches, so much so that he eventually published an extremely circular and nonsensical 'proof' of God's existence to try to placate his attackers.

The irony was that Descartes *knew* God existed – otherwise who had told him to transform philosophy and mathematics via the medium of melons? And, ultimately, as hubristic as this claim was, Descartes did make good on it, publishing the end result of that night in the oven in the 1640s with a slew of philosophical and metaphysical treatises, which were hailed in his beloved Netherlands as 'heretical' and 'contrary to orthodox theology' and 'get out of our goddamn town Descartes.'

Eventually, Descartes found refuge with Christina, Queen of Sweden, who was a fan of his ideas about science and love. She invited him to her court with the promises of setting up a new scientific academy and tutoring her personally. It seemed too good to be true.

It was. In 1649, in the middle of winter, Descartes moved to Queen Christina's cold, draughty Swedish castle and discovered that he couldn't fucking stand his new boss or home. Worst of all for the philosopher who lived his entire life by the principle of never once waking up before noon, Christina declared that she could only be tutored at five in the morning, a demand that Descartes responded to as any night owl would: by saying 'I would literally rather die' and promptly proving his point by literally dying just a few months later. In his final act, the man famous for telling the world 'I think, therefore I am' had posed an equally unknowable philosophical conclusion: he would no longer think, and therefore he no longer existed.

Perhaps the final irony in the tale is that, as heretical as *cogito ergo sum* was considered at the time, with its previously unthinkably radical concept of doubting everything, even that which seems self-evident – modern philosophers have actually critiqued Descartes as *not going far enough*. Thinkers such as Kierkegaard have blasted Descartes for presupposing that 'I' exists at all, and Nietzsche for presupposing that 'thinking' exists.

I guess the moral of Descartes's story, if there is one, is probably this: you can't please all of the people all of the time – especially if they're philosophers. So, honestly? Why *not* just smoke a bunch of weed and crawl into an oven?

7.

Isaac Newton and the Philosopher's Stone

Isaac Newton is probably the most famous scientist to ever have come out of England, which is mostly thanks to some fantastic brand management on his part. That is to say, if you know literally one thing about him, it's almost certainly that he got smacked in the head by an apple and invented gravity as revenge, and the reason you know that story is because he came up with it.

It's a beautifully poetic way to make history, both bringing to mind the biblical creation myth and painting Newton as the kind of person who could come up with the fundamental laws of the universe even while nursing a mild concussion, and that's why he would tell it so often to anybody who would listen, especially if they happened to be writing a biography or something at the time.

In fact, like so many fruit-based events from hundreds of years ago, there's not much evidence it really happened outside of 'he kept on saying it did', but there's no fighting a good story. Today, he's basically reached semi-divinity in his home country, and lest you think I'm exaggerating here, just know that the first draft of his

Isaac Newton and the Philosopher's Stone • 73

epitaph went 'Nature, and Nature's laws lay hid in night. God said, *Let Newton be!* and all was light.'* There are statues, paintings, books and even clipart commemorating the suicidal apple and its historic, if apocryphal, fall from the tree, and Newton himself is constantly making the top ten 'favourite Brits' in those polls that inexplicably keep being run. He even, among the more wry scientific circles, has a whole festival in his honour, with gift giving and good cheer – it's called Newtonmas, and it's held on his birthday, 25 December.

And given how hard he pushed it, Newton would no doubt be very happy to find out how well his story has caught on – especially since it means we're not remembering all the bizarre things he spent the rest of his time doing. There's a reason he's known by some historians as the 'last magician', and it's sure as heck not because they've been won over by the magic and wonder of calculus.

The truth is that, much as we think of Newton as a physicist and scholar today, his life was marked by prophecy from the beginning. He was born in 1642, three months after the death of his father. Much as we make jokes today comparing him to Jesus because of his Christmas birth, at the time things weren't actually that far off – his auspicious birthday really was seen as a sign that he was destined to change the world for the better.

Perhaps it was a surprise for his mum, then, when he turned into a complete and utter psychopath child. It's a common trope that kids don't get on easily with step-parents, but young Isaac took it to an extreme. Maybe it was pent-up rage from all the combination birthday-Christmas presents; maybe it was the fact that his mum pretty much abandoned him when he was three to go marry another man, but Newton went well beyond the usual tantrums and door slamming. Instead he threatened to burn down the family home with his mother, step-dad and half-brother inside.

---

* The final version is even more OTT: it includes the instruction 'Mortals rejoice that there has existed such and so great an ornament of the human race!' And to make it worse, it doesn't even rhyme.

This, ahem, flair for the dramatic would be with him throughout all his many and varied careers: physicist, mathematician, MP, warmonger, eye-poker and detective. But none of those were what you'd call his true vocation – and in fact if you'd asked Newton what he considered himself to be deep down, he'd probably have replied: 'an alchemist'.

Actually, he probably wouldn't have, but that's only because he saw everything he did – the alchemy, the biblical analysis, the ground-breaking physics – as simply different ways of answering the same questions about the nature of the universe. But the fact is that the vast majority of the surviving works of Isaac Newton is not science or maths, but occult studies, and while it was pretty much scrubbed from his reputation for a long time, mostly out of a modern sense of embarrassment, it's true: the man spent a frankly astonishing amount of time researching what was, essentially, magic.

And I'm not using that in a hyperbolic or metaphorical sense – dude was doing stuff straight out of *Harry Potter*. As in, he was literally looking for the Philosopher's Stone.

'Take our matter, put it into a small Retort which is so big as a fist, joyn to it a little receiver of the greatness & form of an egg: close the joynts well with the bladders of an Ox. When they are dry set it in sand & distill first with a most gentle fire. So the matter will become black like pitch & there will ascend into the neck of the Retort a white slimy water which is called & is the Philosophick mercury,' he wrote in an unpublished notebook filled with notes on the Stone.

'It is extreamly volatil & falls down like drops ... strengthen the fire a little & there shall come over a yellow oyle like small veins & drop down into the Receiver, after which when you have made stronger the fire will come a red oyle which is the sulphur of the Philosophers. Continue the destillation with this degree of fire till no more oyle comes over & the Retort becomes wholy glowing: so then will ascend a white fume which is the salt of the Philosophers.'

It doesn't exactly sound very scientific, does it? Neither does his description of the Stone, which is, he says, 'an earth heavy & light being in its nature a watry earth or earthly water. In respect of the colour it is pleasant & abominable, stinking & smelling acceptably. It lies openly & deeply hidden, it is found on hills & dales in fields & streets in gardens & pastures in cellars & shops & yet is found & known by none who is not very wise.'

Perhaps it made sense at the time – although given that he never did manage to turn lead into gold or live for ever (or, indeed, beat Professor Quirrell in a game of wits), it appears the recipe wasn't all it was cracked up to be. Never mind – he spent even more of his time scouring the Bible for details on the Apocalypse, and indulging in what would now be recognised as some Deeply Weird Shit surrounding the Pyramids. To be honest, were he alive today he would *definitely* be in several TV shows and memes claiming that aliens built Atlantis or whatever: one of his pet obsessions was with the Temple of Solomon, which he believed had been built according to a complex mathematical and architectural code that, if only somebody could decipher it, would reveal previously hidden secrets of nature, as if anybody has the time to draw up blueprints that are both structurally sound *and* contain all the mysteries of the universe.

And even when he was actually doing real science, it still looked bonkers. Remember when the Covid-19 pandemic hit, and all those irritating people started telling us to look on the bright side? This might be the most productive year of our lives, they said: after all, they pointed out, Isaac Newton faced two years of quarantine, and he managed to change the world because of it.

And they're not wrong: the time that Newton spent under quarantine at home in rural Lincolnshire, sheltering from the Plague that was ravaging his chosen home cities of London and Cambridge, has come to be thought of today as his *annus mirabilis* – his 'year of wonders'. It's when he formulated what would become calculus and the laws of motion, for example, and it's when he worked out the nature of light – this is when he performed his famous prism experiment that demonstrated how white light could be split into all the colours of the rainbow, a discovery so monumental that it would later be memorialised on a Pink Floyd album cover.

But what all those inspirational titbits about how life under quarantine can spur discovery and innovation tended *not* to point out was the other scientific experiment Newton spent his time carrying out. Which was, to put it bluntly, doing just about anything he could to blind himself.

Have you ever been poked in the eye? It's not nice. You probably saw a few red or blue spots, maybe felt a bit queasy – but either way, you most likely decided to avoid the experience in future.

Newton didn't. Newton saw those red and blue spots and thought to himself, 'time to do a science, and by science, I mean stick a stonking great needle in my eye'.

'I tooke a bodkine' – a bodkin, by the way, being a thick, blunt needle designed for threading ribbons and tape through cloth – '& put it betwixt my eye & [the] bone as neare to [the] backside of my eye as I could', he wrote in his diary next to a helpful diagram of a giant needle being poked into an eyeball. '[P]ressing my eye [with the] end of it … there appeared severall white darke & coloured circles … If [the] experiment were done in a light roome so [that] though my eyes were shut some light would get through their lidds There appeared a greate broade blewish darke circle … Within [which] spot appeared still another blew spot r espetially if I pressed my eye hard & [with] a small pointed bodkin.'

Like all the stupidest decisions in history, it made sense at the time. What you have to understand is that 'science' at this time was really in its infancy, and like all infants, that meant it was basically crawling around in the dirt sticking shit in its mouth and playing with whatever it happened to come across in the hopes of finding something that made it happy. Read the journals from the newly-founded Royal Society in this period, and you won't see cutting-edge research and discoveries – well, OK, you will, but you'll also find things like a letter from Gottfried Leibniz about a sheep he saw the other day that looked like it was wearing a fancy wig, or an account by Robert Southwell of a two-headed cow born in Warwickshire that January.

Get it? *Anything* was science back in these days – they had nothing to go off. So when Newton was sitting there poking shit in his eyes – and this was absolutely not a one-off, by the way; Newton was kind of obsessed with sticking things in his eye, using at various times brass plates and, when all else failed, his own finger to dig around in there – it was because he simply didn't have the distinction between things like 'physics' and 'biology' that we do today. Or, for that matter, the distinction between 'a scientific experiment' and 'permanent injury to the optical nerve'.

So when these attempts failed to maim his sight, he tried something else in his quest to understand how light works. And this experiment was guaranteed to work, as long as your definition of 'work' is 'render a person literally blind for an extended period of time': he was going to stare directly at the Sun.

But, well, any primary-school kid can regret looking at the Sun. That's not *science*. Newton was going to do something *special* – he set up a situation almost perfectly designed to cause as much damage to his eyes as possible.

'I looked a very little while upon ye sun in a looking-glass wth my right eye & then turned my eyes into a dark corner of my chamber & winked to observe the impression made & the circles of colours wch encompassed it & how they decayed by degrees & at last vanished,' he wrote in a letter describing the experience. 'This I repeated a second & a third time. At the third time when the phantasm of light & colours about it were almost vanished, intending my phansy upon them to see their last appearance I found to my amazemt that they began to return & by little & little to become as lively & vivid as when I had newly looked upon ye sun ... in a few hours time I had brought my eys to such a pass that I could look upon no bright object with either eye but I saw ye sun before me, so that I durst neither write nor read but to recover ye use of my eyes shut myself up in my chamber made dark for three days together & used all means to divert my imagination from ye Sun.'

Three days may sound like a long time to be holed up in a dark room nursing your eyes back to health and regretting your life decisions, and it is, but it's actually an understatement on Newton's part. He carried on experiencing these symptoms for 'some months' after this experiment, which is hardly surprising, since the only thing dumber than staring directly at the Sun is to dilate your pupils first by standing in a dark room and staring at it in a mirror.

Possibly the only saving grace of the whole debacle is that he did appear to have learned his lesson: Newton would not mess with the Sun any more. But that didn't mean he was done with his shenanigans.

After all, there were other celestial bodies in the solar system to which he could turn his attention. Hell, he even lived on one. He would set his mind to learning about them – their size, mass, gravitation, everything.

He knew how to do it: he could use all that maths and physics he'd figured out after that apple supposedly fell on him. In fact, the calculations would be the cherry on the cake that was his new book: *Philosophiæ Naturalis Principia Mathematica*, or as it's known in everyday parlance, *Principia*.

Principia is one of those books where, say if aliens came down one day and demanded proof that Earth should not be zapped into space-dust, this would probably be in the 'best of humanity' gift basket that we sent up. It's got gravity in there; Newton's three laws of motion, including fan favourite 'to every action there is an equal and opposite reaction'; an explanation for tides; all the big hits, but none of that is what made it so important. With *Principia*, Newton had done something never really seen before, and he did it so well that it's probably going to sound unbelievable that people managed to do anything at all up until that point.

What could this amazing breakthrough be? It was mechanics – the maths of how the universe works.

I know, I know: more maths. But, I promise, this really is a Big Deal: imagine if you worked at NASA or somewhere, and you were in charge of figuring out whether some asteroid was going to hit Earth or just fly on by peacefully. If *The Martian* and *Hidden Figures* have taught me anything, what you'd probably do is start with some equations describing the asteroid's journey, factoring in Earth's gravity, try to solve for x, give up and make a computer do the hard calculations. But the point is, you'd *start* with the maths.

That *wasn't the case* before Newton. *Principia* contained the first ever proof and explanation of the Laws of Planetary Motion that had been discovered by Johannes Kepler half a century before – all Kepler had been able to do was watch what was going on through a telescope and try to fit some equations with what he saw.

Newton took science from 'the Earth is round, I measured it so I know' to 'the Earth is an oblate spheroid, and I didn't need to measure it because I have these equations which prove it cannot be otherwise'. From 'I dropped a bunch of balls off the top of the Leaning Tower of Pisa and it showed me that all objects fall with the same acceleration' to 'objects fall with the same acceleration, and that acceleration is 9.8 metres per second; we call it gravity and it's related to the size of the planet you're on.'

See what I mean? And really the only problem with it was that he got it wrong, and accidently ended up making the Earth overweight for 300 years.

Here's how you know that Newton's PR campaign worked: he was so revered, for so long, that nobody ever bothered to check his work. It wasn't until 1987, in a lecture hall at the University of Chicago, that anybody spotted the mistake – and as it turned out, it was so simple that an undergraduate student found it while doing their homework.

Look, nobody's saying he should have got it perfect; he was working with some fairly rubbish tools compared to what we have available now, and that meant a particular measurement he needed for the calculation was a little bit out from what we now know it to be. In fact, as far as anybody can tell, it was pretty much just pulled out of his arse.

As a result, he determined our small blue planet to be about twice as massive as it actually is. And given the Earth is about six septillion kilograms – it's around 5,973,600,000,000,000,000,000,000 kg, if you want to see how many zeroes that is – that's quite a lot to be wrong by.

It's a really weird move to make, especially since he was so serious about the rest of his mathematical career that he was willing to start a war over it. Not an actual bloody war or anything – but it does have the distinction of being the only 200-year-long international conflict to be based entirely around mathematical notation.

I am, of course, talking about The Calculus Controversy. Do you remember calculus? You might have studied it right at the end of high school, or maybe later if you carried on doing maths at university – and if you aren't familiar, it's basically the maths of constant change. Acceleration; population growth; radioactive decay; even the spread of viruses; pretty much anything that grows or shrinks or moves in any way relies on calculus to describe it in mathematical terms.

Now, Newton invented calculus at some point in the 1660s, and he deserves some credit for that, but unfortunately he refused to tell anybody about it for about thirty years. By this point, having noticed a gap in the market, Leibniz – remember him? The guy with the wig sheep – had come up with his own version, and, crucially, he actually published it.

Oh well, sucks to be Newton, you might think, but what you're not taking into account is how completely fucking absurd England is and always has been as a nation. Despite it not really making much sense as an idea, what with Leibniz living in an entirely different country and all, Newton loudly accused Leibniz of copying his secret new maths, and for the next couple of centuries scientists from Britain obstinately used Newton's calculus (clumsy but patriotic) while in the rest of Europe everyone used Leibniz's (better but unfortunately German).

But while resisting peer pressure is good for things like smoking or wearing Crocs, it turns out it's not that helpful when it comes to scientific endeavour, and Britain found itself slowly but surely turning into a backwater. They weren't just speaking a different language from their European peers, they were reading an entirely different set of literature: as one anonymous commentator said at the time, British students reading European maths were 'stopped at the first page ... not from the difference of the [calculus] notation ... but from want of knowing the principles and the methods which [continental mathematicians] take for granted as known to every mathematical reader'.

The dispute eventually ended in the early nineteenth century, when a group of mathematicians calling themselves the Analytical Society decided they were fed up of being out-mathsed by their continental colleagues and it was high time to introduce European calculus to the UK. Luckily, Britain learned its lesson, and never again decided to embarrass itself by breaking away from Europe in a doomed attempt to prove superiority over its slightly baffled neighbours. Ahem.

But, as is so often the case in England, pissing off a bunch of French and German folks made Newton even more of a hero than he already was. He rode this popularity right into the House of Commons – though he seems to have been something of a single-issue candidate, since his only officially recorded comments were a request to close the window because he was cold – and was catapulted up to the position of Master of the Royal Mint, where he immediately became a massive finance cop.

No, I'm being serious: sure, if you ask somebody today what Isaac Newton's job was, it's fairly unlikely they'll say 'money detective', but at the time – and for a long time, too; he did this for *thirty years* – he was infamous for exactly that. And to be fair, he kind of saved the national economy from certain doom, which is good – but to be *extra* fair, he sent multiple people to their deaths just for trying to make a little extra in life by exploiting a loophole that the government should never have created in the first place. Which is not good.

See, England had a problem at the time. We're still in the period when most of the coins in circulation were *literally* worth their weight in gold, or at least, worth their weight in silver, because, well, they were made out of silver.

Except, this stupid little loophole was there. Because the coins were *actually* worth a bit less than their weight in silver. As in, if you handed over a silver coin to a silver seller, you wouldn't be able to afford as much silver as you just handed over. Which is about as big a problem as it sounds like, because what are you going to do if you're a poor seventeenth-century fishmonger or hair-powder maker or whatever people did for a job back then? You're going to melt down that solid silver coin and turn it into a lump of metal that's worth more than what you started with.

And since you have the capability to melt down silver, you may as well, you know, take a few little clippings from the coins you have and melt them all down together to make an extra one. I mean, what's the problem, right? It's still the same amount of silver *overall*. Nobody's getting hurt.

But, of course, these early modern life hacks meant that Britain was pretty much haemorrhaging money. And that's where Newton came in: he was put in charge of every coin in the country, and not in an abstract way – he was tasked with *literally recalling every coin in the country* back to the Mint.

It was a mission to which Newton applied himself with aplomb, putting all that mathematical and alchemical nous to use cataloguing and analysing the coins that crossed his desk. This was centuries before the advent of forensics – well before the dawn of any police force, even – and yet here was Newton, a one-man CSI unit, simultaneously reforming the internal workings of the Mint and hunting down counterfeiters with a ruthlessness rarely associated with career mathematicians. He even pioneered the sting operation, sending Mint employees out in disguise to infiltrate criminal circles and bring back intel, and not only did he put his money where his mouth was – or in this case, I suppose, his mouth where his money was – and go out on the missions himself too, but there's records of him personally cross-examining hundreds of witnesses, informers and suspects.

And since forgery, considered high treason at the time, was a capital offence, anybody who got caught up in Newton's investigations would be sent to the hangman. This, as you might expect, made him quite a few enemies: one doomed man referred to him from prison as 'a Rogue and if ever King James came again he would shoot him ... God dam my blood so will I.'

Luckily, King James didn't come back – at least, not in time to stop Newton shaking up the entire economy based mostly on swagger. It was he, with his strikingly modern realisation that Money Is Fake And We Can Do What We Want With It, who first convinced the government to use credit rather than solely hard currency. He talked the country into moving from the silver standard to the gold standard. And just to prove that it really was all reputation rather than skill behind his economical influence, he also invested heavily in the South Sea Company, losing £20,000 – around £5 million today – when the infamous Bubble burst.

If there's one thing we can say about Isaac Newton, it's that he was bloody lucky to be born when he was. Because he lived in a time when 'massive bodies exert a force that pulls smaller things towards them' was about the same level of believable as 'mercury and ox bladders combine to make something that can let you live forever', he's been remembered as the guy who gave the world gravity and calculus in a time when most people still thought dark magic was the number-one crisis facing youth today. He's equally revered to this day by the Royal Mint and the rock band Oasis, both of which have immortalised his famous quote about 'standing on the shoulders of giants' – Oasis by making it the name of a studio album, and the Mint by inscribing it around the edges of the £2 coin. Which is pretty good, considering what the evidence shows: that if he were alive today, he'd most likely be hawking Bored Ape NFTs and spreading conspiracy theories about lizard people on Twitter.

8.

Mozart Uses His Superstar Status to Tell Us All to Kiss His Arse ... Over and Over Again

Even if you know absolutely nothing about classical music, you've heard of Mozart. You can even hum one of his works. That's because he was so accomplished, and so famous, and just so damn *catchy*, that his music is nigh-impossible to avoid these days. Go on: Google 'Symphony No. 40 in G Minor K550' or 'Rondo Alla Turca' and try to stop your toes tapping along. It can't be done.

If you've never seen the 1984 movie *Amadeus*, your idea of who Mozart was as a person is likely based on that one portrait of him you get when you search his name online. He's ... some guy, basically, with a red coat and grey hair and an expression that you only get from truly believing you're several levels of human above anybody else.

Of course, if you *have* seen *Amadeus*, your image of Mozart is probably pretty different, and some might say more accurate. Not, I should be clear, in a factual sense – let's just say if Mozart's teacher and rival Antonio Salieri had been alive in 1984, he'd probably have a reasonable case for libel against the makers of that film – but it definitely got the vibes right: Mozart was loud, unpredictable, and kinda gross, and he was also the creator of a vast musical legacy the likes of which the world has never seen before or since.

But if I had to compare him to a modern figure, I reckon Mozart can basically be thought of as an eighteenth-century Britney Spears. Oh, don't be fooled by that powdered wig – he may have been born in 1756, but his story is one that would absolutely not be out of place in today's celebrity-obsessed world: he was a child star pushed into the limelight way too early by an overbearing father, and like so many such tales it ended in an ignominious family breakdown, a whole load of questionable life choices, and, ultimately, a mysterious and too-early death.

All of which is to say that, if Mozart were alive today, he'd be top of the charts, a constant online shitposter – and as you'll see, I mean that quite literally – and would most likely have some new-age mumbo-jumbo health site on the side that was called something like 'Schmoop' and sold organic placenta tea for newborns. He'd be a superstar.

Which makes sense, since from the very start, Mozart seemed almost perfectly designed by Fate to make history. Well – perhaps we shouldn't give Fate too much of the credit: it was actually his father, Leopold, who was responsible for a lot of it. It was he who had taught the young Wolfgang to play the harpsichord and violin by the tender age of three, and by the time he was six – and already composing his own pieces – Leopold was dragging him and his older sister Nannerl across the continent to showcase them as musical wunderkinds. Between the ages of six and ten, or to put it another way, the age when most kids start primary school and the age when they pick up their first *Goosebumps* books, the young star had been taken from his home in small-town Austria to noble and royal courts in Germany, France, Italy and England, where he, and to a lesser extent Nannerl, were made to perform in front of adoring crowds of European hobnobs.

And as you might expect from the sentence 'these two little kids want to come play the piano for you', the recitals didn't always go exactly to plan – one night in London, for instance, Mozart apparently stopped playing halfway through a piece and ran over to play with the hosts' pet cat. If legend is to be believed, he even 'proposed' to an audience member once: a girl his own age named Maria Antonia Josepha Johanna, from the Habsburg family of Austria, who had caught him when he slipped on stage.

Perhaps she should have accepted, instead of following her family's wishes and marrying the dauphin of France – that union was doomed from the moment she took the name Marie Antoinette – though it can't have been a very tempting offer. First of all, obviously, she was a kid, which would at the very least have created some legal difficulties – but also, and as any working musician will tell you, life on the road really sucks. It sucks now, and it sucked even harder back then, when people didn't even have things like tarmac or suspension, and Mozart's childhood would have been spent bumping about in cold, damp carriages, eating at whatever inns they happened to find along the way, and occasionally recovering from the inevitable sickness that results from trying to juggle being a touring musician with being seven.

And Leopold also worked hard to ensure his son had that other important ingredient for celebrity: severe daddy issues. He showered Mozart with praise, calling him 'the miracle which God let be born in Salzburg', and 'the most amazing Genius, that has appeared in any Age', but equally he, you know, pulled him out of school – or to be more accurate, never let him go at all – in favour of turning him into the family's main breadwinner before he was even old enough to start worrying about his year six SATs. And when Mozart got sick, Leopold's main complaint was that 'this … has cost me 50 ducats.'

And little Mozart was put on the hook for things he couldn't *possibly* have any control over: if the nobles he performed for weren't generous enough for Leopold's liking; if a crowd didn't turn up, or he was too sick to play; later, after Wolfgang went to Paris with his mother and she took ill and died, Leopold's reaction was to blame his son for the death. If only Mozart hadn't been so headstrong and ambitious, he seems to have thought – if only he hadn't tried to branch out on his own, without Leopold running things for him – she might still be alive.

'Your whole intent is to ruin me so you can build your castles in the air,' Leopold wrote to his son, while Wolfgang was still reeling from that loss. 'I hope that, after your mother had to die in Paris, you will not also burden your conscience by expediting the death of your father.'

Leopold even reduced Mozart's official age on tour when he started ageing out of the child genius demographic – presumably working on the logic that it wouldn't be impressive if the tiny child reciting note-perfect renditions of the most challenging music of the day was in fact eight years old and not seven. Really the only thing on tour that Leopold seems to have been a fan of – you know, *really* a fan of, without any little side-complaints or qualifications or accusations of murder – was the flushing toilet he saw in Paris.

We probably shouldn't be surprised, then, that Mozart the adult was ... kind of a handful. He was described by a friend as 'as touchy as gunpowder', and by his patron as 'dreadfully conceited'; he was paranoid and possessive, scolding his wife for going out without him – even to church – and blaming his occasional failures on conspiracies carried out by nameless enemies. And he could be weirdly vindictive, obsessing over Salieri – oh yes: the murder plot was made up by *Amadeus*, sure, but they were contemporaries, and Mozart went hog wild on his rival, accusing him of things like sabotaging Mozart's career and being overly Italian in public. He would even use job applications to dunk on his less-successful peer, telling one potential employer that 'Salieri has never devoted himself to church music.'

And like quite a few modern celebrities, he had some pretty weird and dangerous ideas about health. Like: when his first child was born, Mozart was adamant that the new baby would not be breastfed.

Now, it wouldn't have been that unusual in the mid-1700s for a new mother of means not to breastfeed her baby. It was fairly common

– and often more cost-effective – for the kids to be handed over to a wetnurse for all that, and plus it had the benefit of not forcing you to stay at home with leaky nipples and a baby on your tit all day, which could be inconvenient and gross. If that wasn't an option, there were some attempts at baby formulas you could try: some people advised cows' milk sweetened with sugar, or 'pap', which was small bits of bread soaked in milk and water and sort of pea-shootered down the babies' throats.

I'm not a midwife or obstetrician, and there's statistically a pretty good chance you aren't either, but I don't imagine either of us would be surprised to learn that nourishing a newborn baby the same way you fed the ducks at the park as a kid is not medically advisable. About one in three babies artificially fed like this died before they reached one year old, probably on account of soggy bread being no replacement for mother's milk even when said mother lives in a world where lead-based foundation was seen as a subtle and enchanting day-to-night look instead of a health-and-safety lawsuit waiting to happen.

And yet even these wet nuggets of sandwich would be better than what Mozart had planned for his progeny, which was water. That's all: just water.

'I was quite determined that even if she were able to do so, my wife was never to nurse her child,' he wrote. 'Yet I was equally determined that my child was never to take the milk of a stranger. I wanted the child to be brought up on water, like my sister and myself.'

This, I may remind you, took place in an era when people hadn't yet figured out maxims such as 'don't shit where you eat', or 'the principal risk to health from ingestion of water contaminated with faeces is from diseases such as cholera and other diarrheal diseases, dysenteries, and enteric fevers'. Feeding your baby only water and not milk was basically just giving your baby poison, and nothing else, which is why we can be pretty sure that Mozart almost certainly had *not* been brought up on water as an infant, because of how he was still alive and therefore hadn't died of starvation and E. coli within a few days of being born.

The good news was that everyone who wasn't Mozart knew what a bad idea this was. 'The midwife, my mother-in-law, and most people ... begged and implored me not to allow it,' he recalled, 'if only for the reason that most children ... brought up on water do not survive', which as sole reasons to do things go is a pretty good one.*

And then there was all the bum stuff. Now, traditionally, all Mozart's more quirky behaviours have been sort of glossed over, as simply the price humanity must pay for genius. He was a savant, people thought; *nobody* could be that good and also well adjusted. But modern historians disagree: they argue that Mozart owed his success to innate ability, yes, but also sheer dedication. He worked hard, put tremendous effort into his compositions, and took his job and position in life seriously.

Which would be a lot more believable as a premise if it wasn't for one thing: all the time and energy he spent telling people to kiss his arse.

Yep. It's bad enough being told to 'kiss my arse' by some random bastard on the street, but Mozart was the greatest musical celebrity in the world, and he was damn well going to let you know that.

'Leck mich im Arsch g'schwindi, g'schwindi!' reads the text of Mozart's 1782 canon in B-flat major. 'Leck im Arsch mich g'schwindi! Leck mich, leck mich, g'schwindi!'

It's a pleasant little piece written for six voices, and the lyrics consist solely of the three repeated lines: 'Kiss my arse, quick, quick! Kiss my arse quick!'

Wolfgang Amadè Mozart, it turns out, had an incredibly childish sense of humour – and there was nothing he liked more than a good

---

* He eventually relented in the face of dead babies and common sense, which admittedly puts him quite some distance ahead of, say, modern anti-vaccination advocates.

poop joke. In fact, he was happy enough even if it wasn't good. Or recognisably a joke; the man just liked talking about poop, OK?

'I now wish you a good night, shit in your bed with all your might, sleep with peace on your mind, and try to kiss your own behind,' he wrote to his teenaged love interest in 1777. 'Oh, my arse burns like fire! what on earth is the meaning of this! – maybe muck wants to come out? yes, yes, muck, I know you, see you, taste you – and – what's this – is it possible? Ye Gods!'

That was what he considered romantic. A description of painful constipation. And it was hardly a lone example: we have nearly 400 letters he wrote to various people over the years, and more than one in ten reference poop or butts in some way.

Now, we should address the elephant in the room: Tourette's syndrome. If you've heard of the disorder, you may well be thinking right now that 'Well, this just sounds like the guy had a condition! We can't judge him for that!' And that's fair.

But here's the thing: if you know more about Tourette's than just what you saw in *Not Another Teen Movie*, then you're probably aware that it hardly ever causes coprolalia – that is, the uncontrollable voicing of obscenities (or to translate it directly from the Greek, 'shit-talking'). Coprographia – writing obscenities down uncontrollably – is so rare as to be almost unheard of, and when it does manifest it has this handy loophole where you can cross things out after the fact.

Basically, sure, Mozart may have had Tourette's – quite a few people who knew him did mention what may have been motor tics, like facial grimaces, constant fiddling and fidgeting, and on at least one occasion, halfway through a recital, 'suddenly jump[ing] up, and, in the mad mood which so often came over him, he began to leap over tables and chairs, miaow like a cat, and turn somersaults like an unruly boy', which at least does show that the man was still an unrepentant cat fan. But also, he may not have had Tourette's, and instead just kind of been a bit weird. And even if he *did*, it probably wouldn't have been what caused him to compose entire choral pieces

Mozart Uses His Superstar Status to Tell Us All to Kiss His Arse • 91

in multiple voices, all singing in joyful contrapuntal harmony, about the intricacies of rim jobs.

That's not a joke: in 1786, Mozart turned thirty, and that meant it was time to become mature and sensible. No longer would he write childish bawdy German songs about kissing his arse. No: from now on, he would write them in Latin, like a grown-up.

'Difficile lectu mihi mars et jonicu difficile,' begins Difficile lectu, K. 559, Mozart's canon for three voices in F major. Roughly translated into English, this means ... well, nothing, actually. It's nonsense. But – and here's the punchline – it's nonsense that sounds a bit rude. Go on – say it out loud: 'lectu mihi mars'.

You're not imagining it. Once again, he was just making his performers sing the line 'leck du mich im Arsch' – 'kiss my arse'. And 'jonicu' was reminiscent of the Italian 'cujoni' – think 'cojones'. Basically, the line means nothing in Latin, but it *sounds* like it means 'it's difficult to lick my arse and balls'.

If we want to know why Mozart kept writing jokes about dicks and sex and poop into his work, we don't really need a psychiatry textbook. The reason most likely has a lot more to do with the fact that he was a young guy who had been brought up as a lonely god-child in eighteenth-century Austria – or to put it another way: he just thought it was funny, and nobody had ever told him to behave himself.

'Considered as an artist, [he] had reached the highest stage of development even from his very earliest years,' Nannerl once wrote about her famous brother. But he 'remained to the end of his life completely childish in every other aspect of existence. Never, until he died, did he learn to exercise the most elementary forms of self-control.'

Or take the opinion of Mozart's brother-in-law, Joseph Lange, who described a man who 'either ... intentionally concealed his inner tension behind superficial frivolity, for reasons which could not be fathome', or else 'took delight in throwing into sharp contrast the divine ideas of his music and these sudden outbursts of vulgar platitudes, and in giving himself pleasure by seeming to make fun of himself.'

He was, quite literally, just in it for shits and giggles. And, honestly, it's hard to blame him: the guy only lived until he was thirty-five, and let's just say his sudden decline and death was not exactly *un*expected, since he'd seemingly been falling deathly ill semi-regularly since he was still in nappies – which in Mozart years is probably around the age you write your third or so original composition.

And let's face it: it's not like his work suffered for his arse obsession. Particularly in the last ten years of his life, Mozart was popping out masterpieces like they were eighteenth-century smallpox blisters. It would, you imagine, take a heart of stone to deny him the occasional poop joke.

Which is where Margaret Thatcher comes in. Mozart never met Maggie, which is kind of a shame because he probably would have let rip in ways that *Spitting Image* could only dream of – but then, of course, the two centuries that separated their births would be stretching even Leopold's time-manipulating skills. She did, however, watch a play about him, in 1979: a stage performance of *Amadeus*.

She was *horrified*.

'She was not pleased,' director Peter Hall recalled in an introduction to one printed version of the play. 'In her best headmistress style, she gave me a severe wigging for putting on a play that depicted Mozart as a scatological imp with a love of four-letter words.'

According to the Iron Lady, it was simply 'inconceivable' that the person responsible for 'such exquisite and elegant music' could also be so potty-mouthed. It did not compute – and when Hall pointed out that we have many, many, *many* examples of Mozart being exactly that, she changed tack, and decided she *wouldn't let it* compute.

'I don't think you heard what I said,' she apparently told Hall. 'He *couldn't* have been like that.'

9.

Benjamin Franklin Uses World-Changing Technology to Prank Friends, Self

Few people throughout history have garnered the kind of reputation that Benjamin Franklin enjoys. He could have made it into a book about geniuses for any number of reasons: he was a writer, philosopher, newspaper editor, physicist, diplomat, kite flyer, glasses wearer, Independence declarer, postmaster general, almanac publisher and hundred-dollar-bill model. And had he stopped there – just an average, unremarkable polymath – we wouldn't be talking about him now. But luckily for us, old Benny boy decided to take all that amazing brainpower sitting on top of his shoulders and use it for evil – and by evil, we mean pranking his friends to within an inch of their lives.

Picture the scene: it's summer, 1749, and you've been invited to a party at your good pal Benjamin's house. He's serving a turkey dinner and he's promised all kinds of games after the feast is finished. You sit down, pick up your wine glass to take a sip ... and are hit with an electric shock straight to the mouth.

Unfortunately, this was a typical hazard of being friends with Benjamin Franklin. He had first encountered the wonders of electricity after witnessing an experiment-slash-vaudeville show in 1743, and to say he was electrified by the experience would be a pun too awful to include in this book. He immediately set about recreating – and innovating – tricks in his own home with which he could amaze, and inflict bodily harm upon, his nearest and dearest.

'I never was before engaged in any study that so totally engrossed my attention and my time as this,' he wrote in 1747 to friend and fellow electrician* Peter Collinson. '[W]hat with making experiments when I can be alone, and repeating them to my Friends and Acquaintance, who, from the novelty of the thing, come continually in crouds to see them, I have, during some months past, had little leisure for any thing else.'

Collinson, who lived in England, had just sent Franklin an exciting gift: a brand new, state-of-the-art invention called a Leyden jar. And this device – nowadays relegated to the realms of high-school physics classrooms and *MacGyver* episodes – was quite literally world-changing for Franklin and his contemporaries.

See, historical physicists had been able to *generate* electricity for millennia – in fact, even the word 'electron', Greek for 'amber', comes from the Ancient Greek discovery that you can create a static charge by rubbing pieces of amber. But the Leyden jar allowed scientists to do something nobody had ever seen before: *store* the electricity, and use it on demand.† Gone were the days of being forced to continuously rub long rods of various materials while young boys were suspended from silk above an audience of fancy

---

* This being what they called themselves at the time. Modern-day electricians are, thankfully, generally less likely to maim you with electric shocks just for the lols.

† For the nerds in the audience – the Leyden jar was the first example of what is now called a capacitor.

ladies and gentlemen* – that was now strictly for fun only. With a Leyden jar, Franklin and other serious electricians immediately started 'making electrical experiments, in which we have observed some particular phaenomena that we look upon to be new'.

Leyden jars could not store very much electricity – for comparison, it would take nearly seven trillion of them combined to store the amount of electricity held in a single AA battery today – but for the physicists of the time, a whole new world of experiments and discovery had been opened up.

Now, to give Franklin his dues, he made some seriously groundbreaking discoveries in the field of electronics. He invented both 'battery' the word and batteries the thing† by linking together Leyden jars in parallel, came up with the concept and terminology of 'positive' and 'negative' charges and even the idea of 'charging' and 'discharging' a capacitor. But all of this would come second to Franklin's true electrical vocation: practical jokes. Because while most of us these days hear 'scientist' and picture a serious, white-coated person in a lab,

---

* This was a real experiment – well, more of a parlour trick, really – rather uncreatively called the 'Flying Boy'. It was a speciality of a leading scientist called Stephen Gray, who designed it to illustrate the ability of the human body to conduct electricity. The boy would be connected to a friction generator at his feet – this would be a glass rod, or in some very exciting demonstrations, a mysterious glowing sphere known as a 'sulphur globe', being rubbed by hand against wool or carpeting – and suspended from the ceiling by silk. Once electrified, the boy would be able to perform feats of apparent magic, such as turning pages of a book by simply waving his hands above it.

The grand finale would involve the lights being dimmed and a member of the audience being invited to touch the boy's nose. As their hand approached the boy's face, a loud crack would fill the room and a spark would jump between them, wowing the audience.

Of course, as anybody who has had an electric shock in their life can tell you, those buggers can hurt. This is probably why the flying boys themselves were often street kids, who presumably figured getting suspended from silk ropes and shocked by a few rich dandies was worth it for a few bob. Isn't the past fun?

† OK, this was not a true battery as we would understand it today – that was invented in 1800 by Louis Volta – but there is at least a pretty direct link from Franklin's 'battery' to the AAs in your TV remote.

probably peering intensely at something in a test tube and saying things like 'hmmm, fascinating', back in the eighteenth century the prevailing image of a scientist appears to have been closer to some kind of slightly psychotic circus clown. Not only did his guests have to put up with the aforementioned electrified wine glasses, they were also spooked by dancing spiders made from cork and string, and invited to send sparks of electricity between each other through the medium of air kisses. Franklin even invented a party game, called Treason, in which participants were told to touch a portrait of King George and would receive a shock when they did so, which honestly sounds less like a 'game' and more like one of the warning signs of a serious personality disorder.

'Hold the Picture horizontally by the Top, and place a little moveable gilt Crown on the Kings Head,' he explained to Collinson. 'If now the Picture be moderately electrified, and [a Person] endeavour to take off the Crown, he will receive a terrible Blow and fail in the Attempt.'

'If the Picture were highly charg'd, the Consequence might perhaps be as fatal as that of High Treason,' he wrote, presumably upset that he couldn't literally murder people as a prank without consequences to his party RSVPs. 'The Operator ... feels Nothing of the Shock, and may touch the Crown without Danger, which he pretends is a Test of his Loyalty.'

Not content with the efficiency of singleton electrocutions, Mad Lad Franklin even adapted the 'game' to work with multiple players at once. 'If a Ring of Persons take a Shock among them,' he wrote, I assume while giggling like the villain from a 1940s B-movie, 'the Experiment is called the *Conspiracy*.'

But as the rapper Ice Cube taught us, you've got to chickity-check yourself before you wreck yourself, and there was only so many innocent bystanders Franklin could electrocute before disaster struck. He had been obsessed with the idea of using electricity to cook meat, claiming to anybody who would listen that this made it 'uncommonly tender'. Shooting again for the 'most psychopathic dinner host' award, he would often do this by taking a live chicken or turkey and electrocuting it to death in front of his dinner guests, and it was during one of these live torture-porn displays that he finally met his comeuppance.

'I have lately made an experiment in electricity that I desire never to repeat,' he wrote on Christmas day 1750. 'Two nights ago, being about to kill a Turkey by the Shock from two large Glass Jars, containing as much electrical fire as forty common Phials, I inadvertently took the whole thro' my own Arms & Body, by receiving the fire from the united Top Wires with one hand, while the other held a chain connected with the outsides of both Jars.'

A huge flash and a loud crack filled the room, and Franklin was knocked momentarily senseless. '[T]he first thing I took notice of was a violent, quick shaking of my body,' he recorded, along with numbness and soreness in his body for the next few days.

'I am Ashamed,' he admitted, 'to have been Guilty of so Notorious A Blunder.'

Now, I'd love to be able to tell you that he learned his lesson from this misadventure, but if anything it only made him more psychotic. Just two years later, Franklin embarked on an even more dangerous experiment – one that's now firmly entrenched in American mythology – and it involved a thunderstorm and a key.

Modern scientists will generally advise against going out in a thunderstorm, on account of the many millions of volts of electricity flying around in the sky just looking for a handy conductor to zip through and get to the ground. If you *do* happen to be outside, the best thing to do is to crouch down on the balls of your feet, and put your head between your knees in the classic 'brace, brace' position. Now, I know it sounds embarrassing, but it's true: it makes you as small as possible, and therefore as hard as possible for Zeus to bullseye, and minimises your contact with the ground, making you pointless to aim at in the first place.

I don't think there's an expert in the world, though, who would advise wandering right into the middle of the storm with a kite in one hand and an iron key in the other. And yet, if legend is to be believed, that's exactly what Franklin did one June afternoon in 1752. That was when Franklin, taking his son along with him presumably as some kind of punishment, 'took the opportunity of the first approaching thunder storm to take a walk into a field', according to another electrician friend, Joseph Priestley. The goal, he wrote, was 'to demonstrate, in the completest manner possible, the sameness of the electric fluid with the matter of lightning, Dr. Franklin, astonishing as it must have appeared, contrived actually to bring lightning from the heavens, by means of an electrical kite, which he raised when a storm of thunder was perceived to be coming on.'

And so the experiment began, with key attached to kite string, kite flung into the air in direct defiance of Fate herself, and father and son inexplicably sheltering from the rain in a shed as they waited for either glorious enlightenment or horrific fiery death from above.

Despite the lore – and much to their initial disappointment – the Franklins were not, in fact, struck by lightning on that fateful afternoon, which is good, because it probably would have killed them. But just as the pair were starting to lose hope, Ben noticed something: the hairs on the hemp string keeping the kite aloft were all standing upright.

'Struck with this promising appearance, he immediately presented his knuc[k]le to the key, and (let the reader judge of the exquisite pleasure he must have felt at that moment) the discovery was complete. He perceived a very evident electric spark,' Priestley wrote, possibly mixing up the feelings of 'exquisite pleasure' and 'a sudden intense pain in your hand'. Franklin 'collected electric fire very copiously' in a Leyden jar, and declared the day a success.

It had been a stroke of remarkable luck for the prankster-turned-daredevil: he had diced with death and somehow come out of it alive. But if there's one thing we've learned about Ben Franklin, it's that it's not a party unless somebody gets, at the very least, badly burned by a bolt of electricity.

If that person couldn't be him, and it couldn't be his son, it was going to have to be a stranger. So, naturally, Franklin took to the presses, and sent in a how-to guide to the local *Pennsylvania Gazette*.

'As soon as any of the Thunder Clouds come over the Kite, the pointed Wire will draw the Electric Fire from them, and the Kite, with all the Twine, will be electrified, and the loose Filaments of the Twine will stand out every Way, and be attracted by an approaching Finger,' read his article of 19 October 1752, in what was pretty much the exact opposite of today's 'do not try this at home' disclaimers.

'And when the Rain has wet the Kite and Twine, so that it can conduct the Electric Fire freely, you will find it stream out plentifully from the Key on the Approach of your Knuckle. At this Key the Phial may be charg'd; and from Electric Fire thus obtain'd, Spirits may be kindled, and all the other Electric Experiments be perform'd, which are usually done by the Help of a rubbed Glass Globe or Tube; and thereby the Sameness of the Electric Matter with that of Lightning compleatly demonstrated.'

Exactly what those experiments were, Franklin left to his readers' imaginations – but you have to assume that, if indeed his audience survived the ordeal, they'd be sending a few dinner party invites his way that Christmas. You know, as a thanks for the instructions.

10.

Émilie du Châtelet Cares Not for Your Social Mores, and She Will Fight You in Her Underwear to Prove It

Gabrielle Émilie Le Tonnelier de Breteuil, Marquise du Châtelet, lived her life at full pelt. She was a mathematician, scientist, Newton fangirl, duellist and card shark, and her life stands as testament to the fact that the past sometimes really wasn't that bad a place to be a woman, actually, so long as you had just inhuman amounts of luck, stacks of money, and an unbelievably understanding husband.

Émilie, as she was known, was born in 1706 in France, into a world barely recognisable to you or me. The country was not yet a republic or a democracy; even calling it an autocracy doesn't really do it justice, since France at this point was less 'a country' that operated under 'laws' and more 'a collection of nominally unified provinces' that operated under the rule of 'I inherited the right to forego taxes from my father, so stop arguing, peasant'. Practically nobody was your equal; what rights you held as a subject, along with what rules you followed and even what language you spoke, didn't come to you by virtue of being French, but from your personal mix of attributes, such as occupation, hometown and even surname. And given all this, Émilie couldn't really have picked a better family to be born into. Her father was a minor noble at the court of King Louis XIV who had come by his title not through heredity and tradition but by being basically as French as it's possible to be.* Her mother was a convent-educated woman who knew exactly what was expected from ladies of her and her daughter's rank. The family was rich enough to live in a four-storey house in one of Paris's most exclusive districts, cultured enough to have their own library filled with classic literature and volumes of natural philosophy, and well-connected enough to hold regular *salons*, where the leading thinkers of the Académie Français and Académie Royal des Sciences would come round to eat cake and talk about philosophy and science.

---

* Louis Nicolas Le Tonnelier de Breteuil, Émilie's father, was ... just the worst, to be honest. He was the youngest son in an extremely high-achieving family; his father had worked his way up to one of the most senior offices in the kingdom, and his brothers held important positions in the army, navy, church and the royal court.

Louis Nicolas, on the other hand, was basically the Van Wilder of the seventeenth century, in that he partied hard with daddy's money well past the age where most people would have graduated and settled down, ruined a few women's lives, and would probably be completely forgotten by now if not for the star he bequeathed the world (yes, the part of Émilie's father will be played by Ryan Reynolds in this metaphor). He probably would have got banished from court if it wasn't for the fact that Louis XIV was also, in many ways, just the absolute worst, and the two got on like a *maison en feu*.

At twenty-nine, Louis Nicolas was given the title of 'Reader to the King', a job which had literally zero responsibilities but some of the best benefits available, like respectfully watching the King take a whizz or put his hair on for the day. They both had a penchant for knocking up inappropriate women; Louis Nicolas was sent on several conveniently timed 'diplomatic missions' until the various scandals had died down. Eventually, at the age of forty-nine, he met Gabrielle Anne de Froulay, the twenty-seven-year-old daughter of a well-established family (and technically the sister of his aunt), married her, and had Émilie and her five brothers.

As for Louis XIV, it seems there were no inappropriate women. He was married to Maria Theresa of Spain from the age of twenty-two, but that didn't stop him from taking thirteen mistresses who bore him at least sixteen illegitimate children.

That was just as well, because even the best family in the world couldn't save the young Émilie from her biggest weakness: being a girl. Education for women was just about barely becoming a thing at this point, and most families still agreed with the philosopher Montaigne's judgement that 'a woman was learned enough when she knew how to distinguish between her husband's shirt and his doublet'. It was, to put it bluntly, not seen as the mark of a Proper Woman to have any education or opinions past what we nowadays expect of a six-year-old.

Pretty much the best Émilie could have hoped for was to receive the same education as her mother: instruction in embroidery and elocution at the hands of some of Paris's finest nuns. There is actually evidence that she attended one of these schools, and there's also evidence that she wasn't particularly impressed by it, in the form of her first intellectual project: a translation into French of Bernard Mandeville's banned satirical text *The Fable of the Bees*.

'I feel the full weight of the prejudice which so universally excludes us from the sciences … there is no place where we are brought up to think,' she wrote,* in the introduction to the book. 'I am convinced that many women are either unaware of their talents for lack of education, or that they bury them … for lack of intellectual courage. I experienced this myself, which confirms it.'

Well, Émilie wasn't having any of that in her own life. And so, when it came time for her to be presented at court to find a husband, she hatched a plan. While the other young ladies at court were busy practising the special Versailles tippy-toe walk and curtseying backwards in a corset so painfully restrictive that they had to train for days before they were tough enough to withstand it in public, Émilie was … well, most likely doing the same, in all honesty; that shit was hard, and this was a time when being 'not like the other girls' was

---

* It should be noted that Mandeville did *not* write this. 'Translation' in these days was less about faithfully conveying the meaning of a work originally created in another language and more about just kinda riffing on the general ideas contained inside it, it seems.

more of an accusation than a compliment. But that wasn't *all* she was practising, is the point.

Émilie didn't plan on *just* being seen by the King in an absurd dress and demurely waiting to catch the eye of a posh man. Even at her young age, she knew the realities of life in the eighteenth century: her education, her freedom to climb trees and fence and read books and discuss natural philosophy with the secretary of the French Académie des Sciences – all of that came to her purely through her father's generosity. Most of the other girls her age – even the King's own daughters – couldn't read or write, let alone pen a faithful translation of Virgil from the original Latin into contemporary German.* And so finding the right husband – the man who would take sole ownership of her from her dad – was quite literally a life-changing decision. She needed to make sure she attracted the *right* sort of guy.

So she did what any one of us would do in that situation: she stripped down to her underwear and challenged the head of the King's household guards to a duel.

You can imagine that, to Jacques de Brun, the chief guardsman in the *maison militaire*, this demand for a sword fight from a barely dressed teenage girl must have been rather baffling. It also can't have been a very tempting offer for the seasoned professional soldier: either you lose, in which case what are you doing in your job, really, or else you win and … kill a child?

But for reasons we can only speculate on, de Brun accepted the challenge, and luckily for both parties, it was a draw. At which point, Émilie put down her sword, picked up her copy of Descartes's *Analytic Geometry*, and left.

It was, honestly, a hell of a flex. And it worked: by showing the men at court what a self-confident, accomplished, intellectually curious – all of which was to say: terrible – wife she would be, she successfully bought herself two more years of singledom.

---

* She didn't actually do this, but the thing is that *she totally could have*.

She didn't waste that time. She collected maths books and lab equipment for science experiments, and since her parents wouldn't pay for them she simply learned to count cards and gambled for money to buy them herself.

Eventually, Émilie did marry – she never really had the choice to stay single forever – at the ripe old age of eighteen. These days, getting married at eighteen might make your parents despair at how young you are to be making such a big decision, but Émilie's parents were probably worried about her aging out of the marriage pool, so it probably came as a relief to all involved when the Marquis Florent-Claude du Chastellet-Lomont,* an easy-going army officer who was literally twice her age, showed an interest in her.

Despite the age gap, Florent-Claude was, given the time and place, the perfect match for Émilie. Sure, there were certain wifely duties she was expected to deliver on – and deliver she did, first in 1726 with the birth of Françoise-Gabrielle-Pauline, then a year later when she gave birth to Louis Marie Florent, and finally in 1733 with the birth of Victor-Esprit – but once the family's lineage was secured, Florent-Claude pretty much just went off on military campaigns and left Émilie to do whatever she wanted.

And what Émilie wanted to do was incredibly scandalous. Émilie wanted to study maths.

Now a distinguished lady, she hired some of the leading mathematical minds in the country to tutor her – the kind of people who today have their own theorems and equations named after them. When her teachers met to discuss philosophy and science at the Café Gradot, where women were barred entry unless they were sex workers, she walked straight in after them to join the discussion. And, sure, she was kicked out almost immediately, but you know what she did? She had a tailor knock up a men's suit for herself and started attending the Café in drag.

---

* Basically there were a whole bunch of spelling reforms that happened in the mid-1700s, so when the pair were married they had one surname – Chastellet – but by the end of Émilie's life the standard spelling was Châtelet.

Of course, getting into the maths club at the coffee house is one thing, but actually studying the subject – finding an establishment willing to take her in and instruct her in the latest developments – that would take more than a pair of breeches and a snappy waistcoat. Universities wouldn't take women; that was a non-starter. But there was no rule about hiring all the lecturers to come to her house and teach their lessons there instead.

And once they had, there was no particular regulation against having a torrid affair with them. Perhaps there should have been, since Émilie's extra-curriculars with her tutors drove at least one of them out of a job, and almost certainly not for the reason you think, unless you're into a very specific genre of porn: Émilie's unbound passion literally tired her old schoolmaster out to the point that he had to bow out and nominate a younger, friskier replacement to take his place.

Add to that the near-constant parties, opera and theatre shows, gambling nights, shopping, and everything else a lady was expected to spend her time doing – and we know that du Châtelet was not only willing but eager to fulfil these glamorous obligations – and it's no wonder that modern biographers have wondered when exactly it was that she managed to find time to study.

But study she did – and it wasn't long before Émilie met her two greatest and most enduring loves. The first, a writer and satirist twelve years her senior, was named François-Marie Arouet, and the second was the work of Isaac Newton.

Now, you've probably heard of Isaac Newton – and if you haven't, then please go back a couple of chapters and read up on him, because he was a complete weirdo. Émilie wrote many mathematical and scientific books and treatises throughout her life – she was actually the first woman to ever be published by the Parisian Academy of Sciences – but her magnum opus was to be a French translation of Newton's *Philosophiæ Naturalis Principia Mathematica*. This being Émilie, her version was not only a complete translation, but also included a sort of 'beginner's guide' section, modern updates and commentaries on Newton's theories, and, importantly, a translation of theorems and proofs from Newtonian to continental mathematical notation (remember that Calculus War we mentioned before?).

And I want to be clear here: this was a *massive* undertaking. We tend to forget it now, because of how we're taught about Isaac Newton and the physicist's apple in primary school, but there was a time when the *Principia* was the most cutting-edge physics known to anybody in Europe. Newton's ideas had been out for decades already when Émilie began her work, but there was a reason nobody had tried to do what Émilie was doing. Anybody even attempting such an undertaking would need to be fluent in Latin, geometry, algebra, physics, and calculus – both the English and the continental kind, since the Calculus War was still in full swing; they would need to be up to date on all relevant scientific developments since Newton; they would need to be a good communicator, with an instinct for coming up with intuitive explanations of complex ideas; and most of all, they would need to be actually interested in spreading Newton to Europe. Basically, they would have to be Émilie du Châtelet.

And then there was François-Marie. You've probably heard of him too, believe it or not – but not as François-Marie Arouet. You almost certainly know him by his pen name – Voltaire. He was one of the very few men in Émilie's life who saw her for the genius she was – you know, rather than a sort of The Amazing Sentient Sex Object sideshow act – and their love affair lasted almost the entirety of her adult life.

'No woman was ever more learned than she was,' Voltaire would later write. 'For a long time she moved in circles which did not know her worth and she paid no attention to such ignorance ... I saw her, one day, divide a nine-figure number by nine other figures, in her head, without any help, in the presence of a mathematician unable to keep up with her.'

And like everything Émilie did (and Voltaire did too, for that matter), their affair completely shocked society. Not because Émilie was married – taking a lover was pretty much expected behaviour in those days, since marriage was more about securing family alliances than spending your life with The One. But you were expected to have some *discretion*, goddamnit. You were supposed to *maintain standards*.

Well, if you haven't already guessed, the woman who at sixteen was willing to strip down to her petticoats and challenge a professional fighter to a duel, now married and in charge of her own life, was not interested in the social mores of her prudish peers. She straight-up moved Voltaire into her country home, where they lived for close to two decades essentially as husband and wife, or at the very least as the two main protagonists in a sadly never-produced rom-com sequel in the Ocean's 11 franchise: when the pair needed money, Émilie would go off to the royal courts of Versailles and con it out of the other noble ladies, who, Voltaire wrote, 'playing cards with her in the company of the queen, were far from suspecting that they were sitting next to Newton's commentator'; when Voltaire's gambling debts got so high that he was forced into hiding, Émilie unilaterally invented a new form of what we now know as derivatives trading, buying the right to collect taxes from her posh peers at a discount to raise the money to bail him out. All of which is to say that, yes, if you lost everything in the 2008 crash, you have a loved-up Émilie du Châtelet to blame for at least some of it.

And if you're wondering whether Émilie's husband might have noticed that there was some dude living in his house with his wife, the answer is, yes, and he was totally fine with it. Seriously. In fact, since the house was a bit neglected and Voltaire was spending his own money doing it up, Florent-Claude actively encouraged his living there. Why not, you know? Free renovations!

But all good things come to an end, and Voltaire ultimately fell in love with somebody else – his own niece, as it happens, which is going to become something of a theme with the men in this book. Never destined to simply be a jilted lover in someone else's story, Émilie moved on too, to a toyboy poet named Jean François de Saint-Lambert.*

And then, tragedy struck. She got pregnant.

It was 1749, and Émilie was forty-two years old. In those times, and at her age, she knew that giving birth would be a death sentence. Her translation of Newton had already taken four years of her life, and now she started working eighteen-hour days in a desperate bid to finish it before she died. Voltaire would later write that 'her one thought was to use the little time she thought that remained to complete the work she had undertaken and so cheat death of stealing what she considered was part of herself'.

In September 1749, she gave birth to a daughter, Stanislas-Adélaïde. In quintessential du Châtelet style, the newborn was 'laid on a quarto book of geometry', according to Voltaire, who, along with Florent-Claude and Jean François, was present for the birth. It was kind of like Bridget Jones, I guess, only more so.

Émilie was still making last-minute changes to her translation of Newton right up until she died, six days later, due to complications from childbirth. It took ten years for her work to be published, but to this day it remains the standard French translation of the *Principia*.

In the centuries since her death, Émilie du Châtelet has mostly been remembered as either 'Voltaire's girlfriend', or 'Newton's translator'. Neither does her justice. She was a woman who wanted us to 'be certain of who we want to be'; to 'choose for ourselves our path in

---

* Ms du Châtelet had a knack for picking up lovers that would make Barbara Cartland sit back in admiration. Not only could she count romantic young poets and established rebel philosophers among her exes, but at one point – when, as she put it, Voltaire's 'little machine' wasn't working properly – she also struck up a temporary affair with the son of an honest-to-goodness pirate.

life, and let us try to strew that path with flowers'. She was a scientific genius at a time when it was remarkable that she could even read. And, most of all, she was a scholar, willing to fight for the right to study maths and science.

With a sword, dressed in her underwear if necessary.

II.

Johann Christian Reil Invents Psychiatry and Things Get Really Weird Really Quickly

There's an old medical saying – well, there isn't, but there should be – that goes like this: if you're going to suffer from mental illness, try not to do so before about 1990.

That's because, in the past, people with what we would now recognise as mental health conditions could look forward to some of the most painful, dehumanising treatments known to medicine.* Even in the twentieth century, psychiatric wards – or 'asylums', as they were called at the time – could be more akin to prisons: patients, many of whom had been locked up on the back of vague or exaggerated diagnoses, were stripped of any means of self-expression and personal autonomy and forced to submit to physical restraints, 'treatments' like electroconvulsive therapy† or lobotomies, and in some cases, forced sterilisation.

---

* A term which, here, is being applied *extremely* loosely.
† A quick note: electroconvulsive therapy, or ECT, can be a legitimate treatment and is still used today – but under general anaesthesia, and usually only for those suffering from extreme, life-threatening depression that has not responded to any other kind of therapy. For some people, it can be a genuinely life-saving therapy. Carrie Fisher, for instance, underwent the procedure

And the further back you go, the more wacky and horrifying things get. In the nineteenth century, you could be locked up in an asylum for almost anything: depression, alcoholism, masturbation, and for at least one poor soul housed in California's Patton State Hospital, 'habitual consumption of peppermint candy'. For anybody born without the advantages of being a rich, straight, able-bodied white man, things could get even darker: women were frequently locked up merely for questioning their husbands,* enslaved people were labelled insane for desiring freedom,† and people with chronic disabilities such as epilepsy or traumatic brain injuries were pretty much just thrown into institutions and forgotten about.

Go much further than that, though – back to the eighteenth century and beyond – and the treatment of mental disorders starts

---

regularly (and voluntarily) to treat her own bipolar disorder. 'Some of my memories will never return,' she admitted on *Oprah* in 2011. 'They are lost – along with the crippling feeling of defeat and hopelessness. Not a tremendous price to pay.'

This stands in sharp contrast to how ECT was often used in the past, which was with patients awake and unsedated – a practice that could result in broken bones from the body thrashing about as the current passed through the brain – and often as a punishment or threat for 'unruly' patients. Even when it was being used as a medically prescribed therapy, the 'illnesses' it could be deployed for ranged from schizophrenia to being gay.

* This is not an exaggeration. Elizabeth Packard, for example, was committed against her will in Jacksonville, Illinois, in 1860 without any kind of formal diagnosis, after she decided to change which church she attended. Her husband declared her insane, citing as proof the facts that she was over forty and didn't like that he kept telling everyone she was insane. She was confined for three years, until her oldest son turned twenty-one and was legally allowed to release her; her husband reacted by locking her in the nursery and nailing the windows shut.

The case actually went to trial, and after the jury took a whole seven minutes to declare Elizabeth sane, she was finally allowed to go back to her former life as a local schoolteacher and mother of six. Unfortunately, when she got home after all this, she found that her husband had rented out the house to another family, sold all the furniture, and fled to Massachusetts with her children, personal possessions and money.

† Also not an exaggeration. The 'disease' was called drapetomania, and the 'cure' was whipping.

looking very different indeed. The whole concept of mental illness is different, in fact, on a fundamental level: it's seen as a physical problem, requiring physical remedies.

Unfortunately, physical remedies in The Past tended to range from prayer and repentance (useless, but at least not harmful) to things like ice baths, bleeding, forced laxatives and vomiting, and extensive use of whips and chains (useless and also extremely harmful). Even worse, if you ended up in the wrong asylum, like London's Priory of the New Order of our Lady of Bethlehem, or 'Bedlam', you would have to deal with all this while being gawped at by a bunch of random people who had paid your torturers a penny each to come in and treat you like a zoo exhibit.

Evidently, something changed in between the eighteenth century, when mental illness was either a physical malady or a demon infestation, and the nineteenth and twentieth century, when it was – well, mental illness. So what happened?

Quick crash course on Western philosophy at that point: we're past ancient and medieval thought by the eighteenth century, and well into the period of time known as the Enlightenment. Europe had just been through a couple of centuries of Protestant revolutions, Catholic counter-reformations, religious wars and various Inquisitions, and it was time for a change. A wave of secular, rational thinkers had come to the fore, and philosophy was more empirical and individualistic: the universe was now a collection of physical phenomena, and all that was needed to understand it was careful and thorough scientific enquiry.

Now, you might be wondering what this has to do with psychiatry, and that's fair. But the way a society sees mental illness – along with everything else – is directly tied to how they think about the universe in general. So, in antiquity, people figured mental illnesses were the result of things like head injuries, getting too scared or sad, or being an adolescent, and accordingly they treated them by keeping them at home with their families and trying to talk them out of it, or giving them nice baths and massages to make them feel better. Failing that, there was of course that old favourite of societies everywhere: shunning them and leaving them to wander the streets alone and in distress. Hey, just because it was nicer than you expected doesn't mean it was *nice*.

Then, with the rise of the Church in medieval times, scholars puzzled over whether psychoses were caused by demonic possession or divine inspiration – if the former was deemed responsible, the cure would be an exorcism; if it was the latter, you might end up being venerated as a saint.*

But as the Middle Ages matured into the Renaissance and then the Enlightenment, the world changed. No longer were we living in an ordered system that existed at the behest of an almighty deity; the universe was now a potentially infinite chaos that we were woefully underequipped to understand. Our only hope was to use the one thing that separated us from the animals: our rationality.

In some ways, this would have been a welcome change for those in society suffering from mental illnesses. The change from the old superstitions to new 'enlightened' rationality meant people started saying previously unthinkable things, like 'Hey, maybe these women aren't witches, maybe they're just depressed?' and 'Perhaps it would be good if we didn't just let crazy people starve to death in the streets?' But it also set the stage for some of the worst aspects of how we see mental illness today: not as a mysterious medical problem or a curse by a demon, but as a personal fault, deserving of punishment.

It was into this society, in north-west Germany in 1759, that Johann Christian Reil was born. For the most part, he lived a successful but fairly unremarkable life: he grew up the oldest of five children of a

---

* See, for example, Saint Catherine of Siena (1347–1380), who whipped herself three times daily, was convinced she had invisible stigmata, and had visions – we might call them hallucinations today – of Christ throughout her life. When she was twenty-one, she claimed he had proposed to her with a ring made from his own foreskin, which despite the words you just read, she accepted.

Lutheran pastor, came top of his class in high school before enrolling in university to study medicine. He moved around Germany a lot as a student, studying anatomy under some of the most famous doctors of the day, before eventually moving to the city of Halle and becoming both a practising doctor and medical professor at the local hospital. He wrote books and papers, founded medical journals, and is remembered today in the form of a number of anatomical features and conditions such as the Islands of Reil (in the brain) and Reil's finger (in the finger).

So far, so standard – albeit a bit of an overachiever – and honestly we might not have heard of him today outside of a few highly specialised medical textbooks had he not been born with just enough innate contrariness to take him to Berlin for a year just after he graduated medical school.

Armed with letters of introduction from a freemason named Goldhagen whom he had befriended at university, Reil made his way to the renowned doctor Marcus Herz. Herz was something of a self-made intellectual: born into a very poor family and originally destined for a life doing something or other as a merchant, he had taken himself off to university as a teenager, where he studied philosophy under Immanuel Kant for a bit until he couldn't afford it any more. He went back to work for a while, this time as a secretary to a wealthy Russian, until eventually he came back to Germany and studied medicine in Berlin. Starting in 1777, and ramping up after he married the beautiful and charming* Henriette in 1779, Herz ran one of the most prestigious salons in the city, attended by pretty much all of the leading lights of German intellectual life, as well as quite a few who just wanted to be seen as such, like the Prussian royals.

While in Berlin, then, Reil would have been exposed to cutting-edge ideas in medicine, science and physiology, as well as literature and philosophy. He would have been one of the first people exposed

---

* Also: teenaged. Henriette was fifteen when she married Herz; he was thirty-two. Pretty creepy dude.

to Kant's *Critique of Pure Reason* and the ideas of the French revolution, and the burgeoning Romantic movement being led by poets like Schiller and future patient Goethe.* By the time he moved back to Halle to take up his position at the city hospital, he had added to his natural intelligence and anti-authoritarianism a year's worth of life experience and a deep appreciation for modern science and philosophy.

Put all that together, and you get somebody who's going to change the world. In 1795, Reil, now a highly respected physician and professor, set out on a quest. Medicine, he had astutely noticed, was a mess: barely a science at all, in his opinion, as it had grown bloated over the years with superstition and metaphysical supposition. The problem, he decided, was that the doctors of the day didn't know what they didn't know – that is, nobody had a firm idea on the boundaries of current knowledge. The question of 'life force', for instance – every scientist and philosopher had their own idea of what it was, none of them were exactly the same as any other, and all of them were explained with a fair amount of hand waving and tactical mumbling in order to dissuade any unwelcome probing.

To this problem, Reil applied the philosophy of rational enquiry that had been so firmly established in the European Zeitgeist by this point. To him, the body's 'life force' was a result of all the tiny physical and chemical reactions going on throughout the body, building up the blocks of a human being until eventually you reach the person as a whole.

---

* Johann Wolfgang von Goethe, aka Goethe, was basically Germany's answer to both Shakespeare and Benjamin Franklin in one kidney-stones-afflicted package. He wrote novels, plays, poetry, scientific treatises (real, good ones, not the kind you probably think of when you hear 'Hey, you know the guy who wrote *Faust*? Well, he's just come out with a paper on the nature of colour.') and is generally considered to be the greatest literary figure in modern German history.

Don't feel too bad for him on account of the kidney stones, by the way. He may have been a literary and scientific genius, but like Herz, he was also kind of a creep. When he was seventy-two, he 'fell in love' with a teenage girl he met at a spa town while he was recovering from a heart condition, even proposing to the (just to be clear) child – an invitation she, understandably, declined.

Now, it's probably worth taking a second to note how forward-thinking this actually was. It's probably not that controversial an idea today: we know now that thoughts and emotions are the result of electrical signals in the brain, for instance, or the chemical reactions that take place in our bodies to turn oxygen and water into carbon dioxide. But back then, it was pretty revolutionary: Reil ascribed no 'purpose' or 'duty' to the body like traditional thinkers had, and he critiqued in sometimes scathing terms any physician who took a more supernatural stance.

But if that were the case – if there was no end goal for the human body – and only physical causes, however minute or chemical, of sickness, then medicine should treat people accordingly. And that, ostensibly, was how, in 1802, Johann Christian Reil first got into psychiatry.

Of course, it wasn't called 'psychiatry' at that point, and he initially presented his ideas as pertaining to the treatment of fevers – which, in fairness, absolutely can cause things like hallucinations and mental confusion. But he had another reason to be interested in the subject: he had a number of friends who had suffered and died with various mental illnesses – one of whom was his old friend Goldhagen.

'For a time he was so utterly hypochondriac that the most improbable things terrified him tremendously,' Reil recorded after his friend's final illness. 'He could convince himself of the strangest whims as soon as they came upon him.'

And amid all these personal tragedies, something interesting was happening: across Europe, the Enlightenment-influenced idea that mental illness was the result of a person rejecting rationality was being seriously challenged. Reformers like Philippe Pinel in France, William Tuke in England and Vincenzo Chiarugi in Italy had very recently revolutionised asylums by introducing radical concepts, such as 'speaking to the patients' and 'not keeping them literally chained up', and Pinel in particular was making waves in the scientific world with his push for more humane treatment of the mentally ill.

In 1800, Pinel published his *Medico-philosophical treatise on mental alienation, or mania*. It is rightly regarded as a classic of medical literature today: it not only proposed that mental illness could be treated and even cured through therapeutic treatment, but it contained numerous real-life examples from Pinel's practice – tailors, clockmakers, working women, brought low by circumstance and treated using this new 'moral management'. Reil, despite a deep hatred for the French,[*] was inspired by the work, and in 1803 he wrote his own treatise: *Rhapsodieen über die Anwendung der psychischen Curmethode auf Geisteszerrüttungen*, or 'Rhapsodies on the application of the psychological method of curing mental illnesses'.

Five years later, Reil introduced the term 'Psychiatrie' to the world, and a new branch of medicine was born. He argued in an 118-page treatise for specialist doctors of the mind, trained in university in the discipline of psychiatry. He sought to reduce the stigma of receiving treatment for mental health – something we still struggle with today. And he presented a vast range of potential therapies for various maladies, including sunshine, good diet and exercise, and mental stimulation in the form of books, plays and poetry.

Sounds good, right? So why is Reil in a book about idiots?

Well, in among these therapeutic treatments, consider Reil's suggestion for the treatment of wandering attention: the *Katzenklavier*. Directly translated, that means 'cat piano', which is an extremely accurate description: Reil explained that a collection of cats should 'be arranged in a row with their tails stretched behind them. And a keyboard fitted out with sharpened nails would be set over them.'

Yes, it's as bad as you're imagining.

---

[*] Disclaimer for anybody reading who doesn't have a grounding in European history: this shouldn't be read as something strange or exceptional. You can, at pretty much any point in history, assume any European is filled with a deep hatred of the French, and/or the Germans, and/or the English. Yes, even when they're actively allied with them. It's just tradition.

'The struck cats would provide the sound,' Reil wrote. 'A fugue played on this instrument – when the ill person is so placed that he cannot miss the expressions on their faces and the play of these animals – must bring Lot's wife herself from her fixed state into conscious awareness.'

Of course, Reil knew that the cat-piano wasn't an ideal solution. 'The voice of the jackass is even more heartbreaking,' he acknowledged, but, sadly, he considered donkeys to have too much 'artistic caprice' to be of any psychological use.

Then there's the fact that, along with exercise and diet, he prescribed sex and drugs to treat certain illnesses, directing his patients to drink opium-spiked wine and engage the services of prostitutes to cure their ills. Rock and roll, the third leg of the unholy trinity, would sadly not be invented for another century and a half, which was a shame, because listening to some good tunes may actually have helped some – Reil recognised music as an important therapeutic tool, saying that it 'quiets the storm of the soul, chases away the cloud of gloom, and for a while dampens the uncontrolled tumult of frenzy'. On the other hand, having unprotected sex all over the place was likely spreading syphilis, which would later be discovered to have caused quite a hefty chunk of mental illness at the time.

Then there were the 'hysterical mutes' Reil claimed to have cured by applying strong irritants to the soles of their feet – something you may recognise as basically torture. Or the people he treated with baths – nice baths, ice baths, baths full of eels; all were suggested as potential ways to jolt people out of their 'frenzies'.

And yet, of course, bonkers though this was, it was *still a whole lot better than you could expect almost anywhere else*. Reil may have written prescriptions that sound more like an RSPCA-baiting internet challenge than actual medicine, but at least he *listened* to his patients, and people were known to travel far and wide to see him. In Halle, to this day, you can find a statue, a street, a corner, a mountain, a bathhouse, a pharmacy and a medical centre named in his honour.

Just two years after Reil used the word 'Psychiatrie' for the first time, Halle University had become the home of the world's first psychiatry course. Reil's *Rhapsodies* has been called the most influential work in German psychiatry prior to Freud, and indeed many of his core tenets from the text – that mental illness can strike anybody, for instance, and should not be stigmatised, or the importance of psychiatry-specific training for doctors – still hold true today.

Without Reil, the world might be a very different place for the millions of people who receive mental health treatment today. We have him to thank for the idea that a 'psychiatrist' should be a specialist – a 'precious physician', as he put it – who treats patients for medical issues with scientific methods.

Even if those methods don't include the use of a cat-piano.

12.

Napoleon Bonaparte's Fluffiest Foe

It is famously the case that history is written by the victors, which is how Napoleon Bonaparte managed to be such a noble and virile man of the people and heroic liberator right up until he was defeated at Waterloo, at which point he became a short, angry warmonger who was so impotent that his most famous quotation all these centuries later is still him declining the sexual advances of his wife.

Of course, Napoleon wasn't really short – he was about 1.69m tall, or 5ft 6in, which was around the average height at the time – and he *definitely* never turned Josephine down by telling her 'not tonight'. I mean, quite apart from the fact that he spoke French, and therefore would have said something more like 'pas ce soir, Joséphine', Mr and Mrs Bonaparte actually enjoyed a very passionate love life, which we know because they wrote some *unbelievably* horny letters to each other over the years while Napoleon was away conquering various places. So we can say for sure, for example, that Josephine had a special sex move called 'zig zags', and Napoleon's pet name for her privates was, bafflingly, 'the Baron de Kepen'. Nobody knows the meaning behind either of these facts, so feel free to let your imagination run wild.

It's probably fair to call him a warmonger, though, since he fought more than seventy battles over his military and political life. He lost only eight times, which is undoubtedly an impressive record, and would probably go some way towards rehabilitating his reputation as a reluctant lover – if only one of those defeats hadn't been against a herd of bunnies.

The War of the Fourth Coalition was the fourth of seven attempts in the late eighteenth and early nineteenth centuries by various European powers to band together and stop the French from conquering the entire continent. It ended somewhere between a French victory and a stalemate: technically, Napoleon won, but mostly through paperwork and advantageous timing, which is to say that everybody involved ran out of resources at a time when France happened to be winning. Although Prussia had been profoundly thrashed by this point, Napoleon knew he didn't have a hope in hell of invading Russia;* meanwhile, Tsar Alexander I didn't want to risk being shown up any more than he already had been in case his subjects back in Russia started to think maybe he wouldn't be all that difficult to overthrow after all. And so, as all great powers have done throughout the ages when everybody knows they're screwed but nobody wants to admit it, they opted for diplomacy instead.

History notes that their meeting got off to an auspicious start, as the Tsar greeted Napoleon with the words 'Sire, I hate the English no less than you do' – a sentiment shared by pretty much everybody at some point in history, including the English – and the Treaties of Tilsit were signed on 7 July 1807. Together, the leaders had created an alliance that stretched from Seville to Vladivostok and included quite a few newly founded countries, which had been donated by the Prussian king purely out of the goodness of his own not wanting to die.

---

* This would be something he would famously later forget.

Understandably pleased with himself at having secured lasting peace across Europe (and against Britain) – or, at least, lasting for two years before the War of the Fifth Coalition broke out; it was a pretty belligerent time – Napoleon decided to celebrate. And, like the megalomaniacal Elmer Fudd he was, he decided to celebrate by taking a group of his favourite generals and military commanders to go wabbit huntin'.

Now, rabbits are not ferocious, and neither are they smart. They're not even particularly robust – at least not compared to a bunch of French military officers with firearms. The reason rabbits have managed to survive as a species is because they are really good at one very important thing: hiding.* And that meant that, unless he was willing to let his emperor wander around looking like an idiot for however long it took to win a game of bunny hide-and-seek, Napoleon's Chief of Staff Alexandre Berthier was going to have to adjust the odds slightly in *le petit caporal*'s favour.

And so, when the day came for the grand rabbit-murder spree, Berthier had thousands of bunnies brought to a field in cages, ready to be released to their doom. The men finished lunch, stepped out for the hunt, and the animals were freed.

Presumably, what the hunters expected to happen next was that the rabbits would take one look at their assembled might and run for the hills like a bunch of filthy *rosbifs*. What in *fact* happened next was that the rabbits took one look at their assembled might and thought, 'Excellent, some humans, who most likely have cabbage I imagine', while hopping merrily towards them.

'But how can I tell it or be believed?' the fabulously named Baron Paul Charles François Adrien Henri Dieudonné Thiébault later wrote of the event in his *Memoirs*. 'All those rabbits, which should have tried in vain [to escape] … suddenly collected, first in knots, then in a body; instead of having recourse to a useless flight, they all faced about, and in an instant the whole phalanx flung itself upon Napoleon.'

---

* OK, two things.

Now, I don't care how cute they are in small numbers: a horde of thousands of hungry rabbits running at you en masse would probably frighten you, even if you're Napoleon and your own brother is a rabbit.* Also, it's kind of hard to brag about what a master hunter you are when you're going up against a prey that is literally running towards you hoping for some snuggles and snacks. So, hoping to avoid making Napoleon look like the kind of psycho who would gather together a group of battle-hardened generals to watch him bloodthirstily mow down a bunch of friendly fluffles, Berthier tried to make the bunnies scatter, ordering the generals' coachmen to drive them away using their whips. At this point the bunnies got the memo and sensibly fled, and Napoleon's party relaxed, relieved to once again believe themselves the alpha in the situation.

Unfortunately for the generals, they had underestimated the enemy. The bunnies, as it turned out, were not so much 'fleeing in fear' as they were 'mounting a tactical retreat to ready battle formations'. As Baron Thiébault recalled: '[T]he intrepid rabbits turned the Emperor's flank, attacked him frantically in the rear, refused to quit their hold, piled themselves up between his legs till they made him stagger, and forced the conqueror of conquerors, fairly exhausted, to retreat and leave them in possession of the field.'

---

* So, Napoleon Bonaparte's brother, who was confusingly named Louis Napoleon Bonaparte, was given Holland when his brother conquered it in 1806. Although basically just a puppet king put in place to ask how high when Napoleon the elder said 'jump', he seems to have been really fond of his kingdom, and as good a head of state as you can be when you're an impotent monarch installed by a megalomaniacal sibling in his quest to consolidate control over an entire continent.

Anyway, in his efforts to be a good guy, he decided to learn Dutch, which was probably a good idea given his kingdom was full of Dutch people. In an effort to introduce himself to his new subjects, he announced in a speech, 'Iek ben Konijn van Olland', which Dutch speakers may note does *not* mean 'I am the King of Holland' but in fact means 'I am the rabbit of Holland'.

Evidently the people of Holland didn't hold this confession of rabbithood against Louis: he is remembered in history books as 'Louis the Good', which is pretty complimentary as regnal nicknames go.

In short, the bunnies had out-Napoleoned Napoleon. They split into two wings and converged upon their supposed hunters in a classic pincer movement, forcing the generals into a rout and claiming the field for themselves.

So how had the generals been outsmarted by a bunch of rabbits? Well, it turned out that Berthier, evidently not a bunnyologist (technical term) had kind of just assumed that all rabbits were basically the same as each other when he went out to buy them for the hunt. Now I'm not going to rhapsodise over the nuances of bunny personalities and foibles, but it's definitely true that *pet* rabbits and *wild* rabbits will react differently upon being faced with a group of humans who smell vaguely of picnic.

'[The rabbits had been bought] from the hutch instead of from the warren,' wrote Thiébault. 'The consequence was that the poor rabbits had taken the sportsmen, including the Emperor, for the purveyors of their daily cabbage, and had flung themselves on them with all the more eagerness that they had not been fed that day.'

The bunnies swarmed the coaches, overpowering the generals with fluffiness. Napoleon jumped into his carriage and beat a hasty retreat, chucking buck-teethed fuzzies out of his path (and his coach) as he left. Fresh from his victory against the combined armies of Britain, Russia, Prussia, Sweden, Saxony and Sicily, the Emperor of France had finally been defeated.

It was over. The bunnies had won.

13.

Lord Byron, the Patron Saint of Fuckboys

You know this guy. He's charming. Cocky. Most likely a bit messed up and insecure, deep down. He knows *exactly* what to say to win over the object of his affection – but once they're hooked, he'll move right on to the next paramour. He's good-looking, sure, but that's not enough to explain the long, long list of jilted ex-lovers he leaves in his wake. The truth is, you know he's bad news – but maybe, *just maybe*, you could be the one to finally fix him.

Recognise the description? Then congratulations: you've met The Fuckboy™. History and culture are replete with examples of the breed – Henry VIII was a prime example, as was Romeo of *Romeo and Juliet* fame* – but for the purest incarnation of the phenomenon, there's really only one name you need to know: Lord Byron.

Now, George Gordon Byron, 6th Baron Byron, didn't have the best start in life. I mean, sure, the fact that we basically think of his first name as 'Lord' these days might suggest the appearance of a silver spoon or two in the delivery room, and, yes, the building he

---

* 'But he was faithful to Juliet! He married her!' Yeah, yeah, go talk to Rosaline and tell her what a noble gent he was.

was born in may now be home to London's flagship branch of the John Lewis department store. But on the other hand, his dad was a man known to history as 'Mad Jack Byron', which doesn't exactly suggest the most stable of childhoods for the future poet. Indeed, within the first three years of Lord Byron's life his father had, in short succession, squandered the entirety of his wife's fortune on booze, abandoned them both, fled to France to start a torrid affair with his own sister, and then died at the grand old age of thirty-five. This latter move was, Byron's friend and biographer John Galt would later write, 'greatly to [his wife's] relief, and the gratification of all who were connected with him' – though probably not to the tiny Byron, who was named in Jack's will as the inheritor of all his scoundrel father's debts.

His mother, meanwhile, was a Scottish heiress named Catherine Gordon of Gight, whose father and grandfather had both died by suicide, and while it's definitely fair to say she was the better parent of the two, she probably wasn't what you'd call a calming influence in young Geordie's life.* Part of that was undoubtedly Mad Jack's fault – it's hard to be mother of the year when random bailiffs keep coming to the door and demanding payment for eighteen casks of gin you don't remember buying – but even without his input, she was known for her wild and violent tempers. She drank heavily, and would attack Byron both physically and verbally – she was a rather large and clumsy woman, and couldn't often catch him, but Byron suffered from an untreated club foot and she wasn't above using his

---

* Catherine was said to have cried out with sadness so loudly that the whole street heard when she learned of her husband's premature death, which is odd because, by all accounts, the pair couldn't stand each other. Jack was rude and violent and terrible with money – within two years of marriage he had spent the entirety of Catherine's £23,500 inheritance, a sum which is worth about £2 million today. Catherine, for her part, was moody, ill-educated, and, most damningly of all for an eighteenth-century woman, ugly. She was Mad Jack's second wife after he had essentially ruined the life of a young (and by that point dead) marchioness a few years earlier, which really should have set off alarm bells for young Cathy.

disability against him. His diaries describe at least one occasion where she referred to him as a 'lame brat'.

Add to that a nanny who would alternate between beating him until, in his own words, his 'bones sometimes ached from it', and sexually abusing him, and you start to figure Byron's 'mad, bad, and dangerous to know' reputation was really something of a foregone conclusion.

'It is impossible to reflect on the boyhood of Byron without sorrow. There is not one point in it all which could, otherwise than with pain, have affected a young mind of sensibility,' Galt would later recall. 'The riper years of one so truly the nursling of pride, poverty, and pain, could only be inconsistent, wild and impassioned, even had his natural temperament been moderate and well-disciplined ... It would have been more wonderful had he proved an amiable and well-conducted man, than the questionable and extraordinary being who has alike provoked the malice and interested the admiration of the world.'

That was written in 1830. Do you know how fucked up your childhood had to be to make the Georgians put down the callipers and embrace attachment theory?

As insufferable and bad at judging nannies as his mum might have been, she did at least seem to love Byron, and when they got too impoverished to stay in London she took him up to her homeland of Aberdeen, rented a flat in Broad Street for the two of them, and made sure he got an education at the local grammar school. He even fell in love for the first time up in Scotland – with his cousin, which just goes to show that heredity has a lot more to answer for than we might assume.

So at this point we have a kid whose formative years have been spent moving from place to place, house to house, with an unstable mum, an absent father and an abusive nanny; who has, since birth, lived with a club foot that's left him in constant pain and with a chip on his shoulder so large it's basically a whole potato; whose family on both sides has a history of what we would now recognise as just ludicrous amounts of mental illness; and to top it all off, he's living in Aberdeen, a city which was as cold and gloomy as it was filled with feuding clans and Calvinist damnation.

And even at this young age Byron was intensely charismatic – which was lucky, because he was also the kind of total brat who probably would have been sold in some kind of Georgian boy-auction like a prototype Oliver Twist if he couldn't charm his way out of trouble. There are stories of him dressing a pillow in his clothes and tossing it out of the window to make his mum think he'd killed himself, for example, or, while staying with a family friend, pulling a loaded gun from the mantlepiece and shooting at the cook.*

Despite all these stories of lordly exuberance, though, it's weird to think how close Lord Byron came to simply being Geordie Byron, Aberdeen lad. At this point in his life, though, that was all he expected to be – so it came as a surprise when, aged just six and a half, he and his mother were informed that he was the heir to the Byron title and estate through the death of his great-uncle, William Byron.

For Catherine and George, in their poky Aberdeen flat, this must have seemed like a miracle. Until they read the fine print, that is: see, the 5th Baron Byron had been a guy known as the 'wicked' or 'devil' Lord Byron, and he was a violent recluse famous for killing his neighbour in a drunken pub argument over whose lands had more wildlife – in other words, a classic of the Byron genre. For the previous couple of decades, everyone had assumed his land and titles would go to his son William, but he had kind of written himself out of the inheritance by getting cannoned to death in some continental war, as was the fashion at the time. Having now disappointed his father as an heir twice over – first by marrying some penniless cousin† rather than a

---

* He missed, and got her hat instead; 'the woman's cap was riddled with the shot', recalled one of the sons of the family Geordie was staying with at the time. 'For this act,' he added, 'he was very properly horsewhipped by Mr Hanson.'
† The Byron family really was vying with the Ptolemies and the Habsburgs for 'family most committed to concentrating the gene pool'.

wealthy heiress, and then by dying – William's promised fortune was sold off and left to ruin, and when George and his mother eventually moved down to the Byron estate in Nottinghamshire, they found the house decrepit and the money tied up in fraudulent debts racked up by the previous lord.

Nevertheless, Byron was now an aristocrat, and that meant he was free to be a wildly unstable lothario-poet-*bon vivant* without worrying about things like not being able to afford tomorrow's haggis. He immediately fell madly in love with two more cousins, started writing poetry, and was sent off to Harrow to play cricket and learn how to be posh. His mother showed the same financial restraint with him as she had his father, which is to say absolutely none; nevertheless, he proclaimed upon leaving his family home that 'no captive Negro, or Prisoner of war, ever looked forward to their *emancipation*, and return to Liberty, with more Joy, and with more lingering expectation, than I do to my escape from maternal bondage', which is, if nothing else, an *exceptionally* rich-white-boy thing to say.

Harrow was, for the most part, not a pleasant experience for the young Byron. After all, to many of his peers, he was a backwards yokel with weird eyes* and a bum leg who couldn't even speak Latin, and as far as his teachers were concerned – well, I'll just let you take a second to reread the bit where he literally shot at a friend's cook – not even his own cook! – over a mild disagreement about dinner. This was *not* a student who responded well to authority.

But it also seems to have been the place where the teenage lord first started having relationships with people who weren't blood relatives: there was a brief but intense summer fling with Lord Grey de Ruthyn, as well as classroom romances with several other boys at school.

And as much as he disliked much of his experience there, Harrow clearly did its job: it took a young, pudgy kid with an estate in the shires and a history of trauma and it spat out a sexually awakened

---

* Apparently they called him 'eighteen pence' because the size difference between his eyes was like that of a shilling and a sixpence.

poet lord with an eating disorder and a hatred of the classics. The stage was set for George Byron to go full wastrel.

Byron went on to Cambridge, taking with him as a pet a fully-grown actual bear in protest at the college's 'no dogs' rule and a reputation, as he boasted to his friend Edward Noel Long, for 'seducing no less than 14 Damsels including my mother's *maids* besides sundry Matrons and Widows'. He fell in love, again, with a chorister named John Edleston, who would later become the subject of several saucy albeit plausibly deniable* poems, and he got really into swimming and boxing in a manic attempt to prove he truly wasn't affected by his club foot, *actually*, in fact, he barely noticed it.

And then, when he and his friends matriculated and all his college pals got grown-up jobs and took on adult responsibilities, he did exactly what you would expect given everything we've seen up until this point: he transformed into a little bitch and called them all posers before fucking off on a gap year and finding himself a new boyfriend named Nicolas Giraud.

'I have really no friends in the world', he wrote in an 1810 letter to his presumably slightly insulted pal Francis Hodgson. 'All my old school-companions are gone forth into that world, and walk about there in monstrous disguises, in the garb of guardsmen, lawyers, parsons, fine gentlemen, and such other masquerade dresses.'

Honestly, being Byron's friend must have been exhausting. If you worked for a living you were a hypocrite, while the aristocracy were little more than inbred sheep – he wrote in *Don Juan* that the upper classes are famous for 'marrying their cousins, nay, their aunts and nieces, which always spoils the breed, if it increases', which, while true, is quite the self-own from George 'Three Cousins' Byron. He

---

* While we can be pretty sure that Byron had gay relationships throughout his lifetime, he and his many boyfriends would have been very aware of the dangers of being open about it. Men were executed for sodomy in the early 1800s at a higher rate than ever before; in 1806, when Byron was eighteen, more people were hanged for sodomy in England than for murder.

refused to seek treatment for his foot, but was obsessed with hiding and overcoming it; he lived beyond his means to the point that his old mum basically lived in hiding from his creditors. He was also endlessly self-hating, moody, and so obsessed with his weight that we'd probably call him anorexic today. And fellow poets, with whom you'd think he would have some level of camaraderie, were frequently the subject of his ire – especially if they had committed the cardinal sin of becoming more successful than Byron himself.

There was Robert Southey, for instance: a former Romantic radical like Byron, whom Byron professed to hate more than anybody else in the world apart from his own mother. When Southey was made poet laureate, Byron dedicated three verses of *Don Juan* to him, saying, 'Bob Southey! You're a poet – Poet-laureate … 'tis true you turned out a Tory at Last', and then implied that his dick didn't work, which is the kind of professionalism that might explain why Southey got the top gig and Byron didn't, if we're honest.

Then there was William Wordsworth, whom Byron called 'Turdsworth' in a testament to the benefits of the English public school education system, and John Keats – 'Jack Keats or Ketch or whatever his names are', Byron wrote – whose 'piss-a-bed poetry' was mere 'onanism – something like the pleasure an Italian fiddler extracted out of being suspended daily by a Street Walker in Drury Lane'.

He used poetry to vent his radical political views too: he wrote fiery verses in support of the Luddite rebellions, and odes cursing Lord Elgin, who was at that point busy stealing some marbles from the Acropolis. And when Lord Castlereagh, one of the leaders responsible for the vicious suppression of the Irish uprising of 1798, died in 1822, Byron composed a heartfelt elegy in his memory:

> Posterity will ne'er survey
> A nobler grave than this.
> Here lie the bones of Castlereagh:
> Stop, traveller, and piss.

Byron has been known to many historians as the 'first modern celebrity', and frankly it's clear to see why. It wasn't just his public smackdowns of his rivals or his obsession over weight loss and fashion; we can easily imagine a modern Byron obsessing over his Instagram and Tinder accounts, or topping various magazines' Sexiest Man Alive lists. Even the Duchess of Devonshire, herself a celebrity of played-by-Keira-Knightley stature, said that 'the subject of conversation at the moment is not Spain or Portugal, warriors or patriots, but Lord Byron. His poem is on every table, and he is courted, flattered, and praised wherever he appears.'

And boy, did he enjoy it. Byron spent his fame drinking, partying and racking up notches on his bedpost. Then, evidently not aware of the life maxim 'don't stick your dick in crazy' – or, more likely, aware of it but seeing it as a challenge rather than a warning – Byron started his affair with Caroline Lamb.

From a modern perspective, it's hard not to feel sorry for Caro, as Byron called her. She was naturally super smart and gorgeous, but thanks to her unbelievably shitty upbringing she had barely any formal education and what we might think of today as bipolar disorder. She was already married, but the stillbirth of one child and the severe mental disabilities of another had caused a kind of mental breakdown in the twenty-six-year-old Caroline, and basically what this means is that meeting Lord Byron in 1812 was quite possibly the worst thing that could have happened to her.

In fact, it was Caroline who coined the now famous description of Byron: 'mad, bad, and dangerous to know'. And she was speaking from experience: the pair were only together for a matter of months, but it was wildly passionate and scandalous. Caroline would dress as a pageboy and sneak into Byron's room for sexual rendezvous, and there was talk of running off together to make a new life in exile. Caroline did things like sending Byron a lock of her bloody pubic hair as a keepsake – 'I cut the hair too close and bled more than you need', she explained in the attached note – and immediately smashing a glass and using the pieces to try to slash her own wrists when Byron insulted her at a party.

Which, by the way, was part of the problem – because after taking on a married woman with clear mental illness as a lover, Byron was then unfailingly a gigantic arsehole to her. He eventually dumped her with a note that just said, 'I am no longer y[ou]r lover', sending her into a depressive episode so severe that her husband moved the family to Ireland as a form of rehab, leaving Byron free to immediately knock up his own half-sister. Like father like son, I guess.*

Then, to really fuck her over even more, Byron married Caroline's cousin, Annabella Milbanke. As you can imagine, this would have made birthdays and Christmases extremely tense, but the heart wants what it wants, right? Except that the heart most emphatically did not want this union – neither Byron's nor Annabella's. In fact, to give you an idea of just how serious he was about the relationship from the get-go, Byron never actually proposed to his purported sweetheart, instead preferring to just finagle and manipulate circumstances so that she had no choice but to marry him. His excuse was that he needed her as a sobering influence, which is – little piece of life advice for you here – one of the absolute worst reasons to marry somebody, and especially a person who has repeatedly called you her 'friend' and said things like 'I'm so glad you're my friend', and who, by the way, you've barely ever actually met in person.

So, unsurprisingly, the marriage was a match made in hell: he thought she was prissy and detached, and she thought he was insane. The marriage lasted just long enough to produce a daughter, Ada, before Annabella got so sick of Byron's philandering around with various actresses and half-sisters that she took Ada to her parents' and

---

* Caro did get some Georgian nobility-level catharsis eventually: she burned all the gifts and letters Byron had given her in a huge public bonfire on her husband's estate while the village children danced around the flames and a page recited a poem about what a dickhead Byron was and how he never deserved her anyway. We can all only hope to send off our exes with such pageantry.

sought legal advice for a formal separation. The marriage lasted little more than a year in total.

By this point Byron was in one of his Moods again, saying that he 'should, many a good day, have blown my brains out but for the recollection that it would have given pleasure to my mother-in-law'. Fed up with England and its increasing swarms of debt collectors and spurned ex-lovers, he decided to leave the country entirely: he travelled to Switzerland with the poet Percy Bysshe Shelley and his novelist wife Mary* and got yet another inappropriate woman pregnant. This time, it was Claire Clairmont, Mary's younger sister – the resulting daughter, Allegra, was sent to a nunnery where Byron never once visited her.

Then, in Italy, he met the Countess Teresa Guiccioli, a woman twelve years younger than him but more than fifty younger than the man she had married three days earlier. Not only did the pair start an affair that lasted more than five years – which actually places her pretty high in terms of length of relationship with the exhaustingly flighty Byron – but they carried it out very openly, even moving Byron into Guiccioli's palace. Guiccioli himself, apparently an extremely understanding count, only evicted Byron when he found out the lord had been stockpiling weapons in his room to aid the local revolutionaries in their bid to overthrow the nobility, Guiccioli included. Which, you have to admit, is kind of a baller move on Byron's part.

And throughout this time, Byron was working on his magnum opus: *Don Juan* – and that's 'Joo-an', not 'Hwan', by the way, because as anybody who's seen English people on holiday can attest, they're bloody terrible at blending in and Byron by this point was done being subtle. It was a fucking massive poem – 16,000 lines long – and completely scandalous. We've already touched on the dedication that he used basically as a callout post against other writers, but the poem itself, too, was nothing more than pure, distilled Byron: it followed

---

* Yes, she of *Frankenstein* fame; this was in fact the very summer getaway that led to her writing the novel.

the adventures of a suspiciously Byronic figure who was full of wry observations along the lines of 'don't spend too long away from your wives or else somebody, not saying who, might come and sex them up'.* He was scathing, ridiculing society and individuals – and he paid for it, too: reviews called him immoral, with *Blackwood's Magazine* saying 'this miserable man ... [has] drained the cup of sin even to its bitterest dregs ... he is no longer a human being ... but a cool unconcerned fiend, laughing with a detestable glee over the whole of the better and worse elements of which human life is composed – treating well-nigh with equal derision the most pure of virtues, and the most odious of vices – dead alike to the beauty of the one, and the deformity of the other – a mere heartless despiser ... whose type was never exhibited in a shape of more deplorable degradation than in his own contemptuously distinct delineation of himself', which is ... pretty intense, as book reviews go.

Byron, for his part, thought the whole thing was hilarious, writing to his publisher that 'you ask me for the plan of Donny Johnny; I have no plan – I had no plan; but I had or have materials ... You are too earnest and eager about a work never intended to be serious. Do you suppose that I could have any intention but to giggle and make giggle? – a playful satire, with as little poetry as could be helped, was what I meant.'

Then, in 1824, at the age of thirty-six – yes, seriously, all this scandal and drama happened before Byron made it to thirty-six – he was invited by Greek revolutionaries to help them overthrow the Ottoman Empire. Naturally, therefore, he immediately sold his estate and sailed to Greece, where he adopted a child, fell into unrequited love with a boy named Lukas Chalandritsanos, and spent vast amounts of money single-handedly refitting the Greek navy. He went to Missolonghi with a 200-some strong battalion, again paid for entirely out of his own pocket, where he was meant to lead

---

* Or in his words, 'for all that keep not too long away; I've known the absent wrong'd four times a day.'

an attack against the Turks despite the fact that his entire military experience could be summed up as 'accidentally inherited a title; is rich'. Instead of leading the charge, however, he almost immediately got typhoid, and despite his doctors doing everything they could for him, which in this period is code for draining just absurd amounts of his blood, he died.

Although he had made it clear that he wanted to be buried in Greece, his body was taken back to England, where his body was refused a plot in Westminster due to his 'questionable morality'. No doubt he would have approved; the writer William Hazlitt even went so far as to say that if he *had* been interred there, his body would probably get up and walk straight back out again.

Since his ridiculous life and death, Byron has become many things. In Greece, he was a hero; in England, he spawned a whole genre of literature and became the closest thing the Brits ever had to a Casanova. He was a poet, a lover, an adventurer, a sportsman, a bi icon and a bear owner – but most of all, he was a messy, petty little bitch who couldn't keep his pants on. And for all that he wrote a good rhyme or two, one thing's for sure: the Georgians were damn lucky they didn't have Twitter back then. Or, for that matter, Grindr.

14.

Ada Lovelace's (Husband's) Family Jewels

It may not be fair, but there's undeniably a certain ... stereotype associated with computer programmers. There's Frink from *The Simpsons*, Dexter from *Dexter's Lab*, heck even Bruce Banner when he isn't in one of his big green moods – they tend to be awkward, reclusive and, of course, men.

So if somebody asked you to imagine the world's first computer programmer – the Ur-programmer, if you will – maybe your first thought would be to just turn this character up to eleven. Make the glasses extra thick, pull the pants up to the armpits. Make it, like, 1953. Maybe his name is Eugene.

Yeah, you couldn't be more wrong. The world's first computer programmer was born in 1815, named Augusta Ada King, and was kind of a party girl.

Let's take a step back – remember that brief mention of a daughter in the previous chapter? OK, it was Byron, so we need to be more specific: remember that brief mention of the only one of Byron's daughters born within wedlock? This is her.

Which, honestly, makes it all the more impressive that we've heard of her at all. Ada, as she was known throughout her life, could easily have become just some nineteenth-century 'it girl' who fell into fame thanks to her notorious aristocratic dad. Instead, she forged her own path, and became just as much of a fucking mess as her father before her.

Now, as you may recall, Lord Byron and Annabella Milbanke were just the absolute worst match for each other. God knows what first attracted the pair together, but it didn't last, and that's why Ada was taken by her mother to live in Leicestershire when she was just a month old and never even saw a picture of Byron until she was twenty. In fact, so monumentally pissed off was Annabella that she moulded Ada's entire life around her anger, treating Ada sometimes like an unwelcome continuation of her ex-husband and other times like a thematic device that you'd blast for being too unsubtle if a director used it in a movie about overcoming a bad ex.

And most of the time, she simply wasn't there. Ada was basically left with her grandmother, Judith, for most of her childhood, and had just her pet cat Mrs Puff for company. Annabella would write anxious letters to Judith asking how Ada was getting on, which sounds like a nice thing until you learn that the letters came with cover notes asking Judith to keep them as proof that she gave enough of a shit about her daughter to prevent Byron getting custody. She occasionally let the mask slip: in one letter she told Judith that 'I talk to it for your satisfaction, not my own, and shall be very glad when you have it under your own.'

Just to be clear, the 'it' she's talking about in that quote was Ada.

So, from a young age, Ada showed an aptitude for two things. The first was science, and that was at least partly by design: her mother, working on the assumption that she could cancel out bad Byron genes with a good Milbanke education, directed her to study logic and maths and self-control rather than dangerous things like poetry.

But even without that influence – or straitjacket, depending on how you look at it – Ada seemed to have inherited some of her mother's scientific curiosity. When she was twelve, for example, she got obsessed with what she called 'flyology', methodically investigating and sorting through various materials, construction methods, anatomical requirements, possible power sources – all the different kinds of things a girl would need to know if she wanted to take to the skies.

The second thing Ada was good at was falling sick with mysterious illnesses that could be treated only with mathematics and opium. This was almost certainly because it was the 1800s, and what else are you going to prescribe a *woman* who *complains*, but it also may have had something to do with the fact that she had grown up under the eyes of a mother who called her 'it' in correspondence and would send people to spy on her rather than actually visit. And while I'm sure, detached as she was, Annabella would have said she was looking out for Ada's best interests by doing that, the fact is that she wasn't interested in how happy or healthy Ada was so much as how morally upright – basically, how *not like her father* – she was acting.

But if Annabella had hoped that forcing her daughter into science and maths would neutralise any Byronic tendencies, she would be very disappointed. In fact, all she managed to do was sort of ... *redirect* the chaos. Ada certainly did become a brilliant mathematician, and a visionary a hundred years ahead of her time when it came to computing, but she was, without a doubt, her father's daughter – which is how she still ended up scandalising society with her attitudes to sex and monogamy, her smoking and swearing, and why she ended up thousands of pounds in debt to the bookies, her friends and her husband after trying to use her maths skills to win big on the horses.

It all started, as things do when you're a fancy lady in the 1800s, with a coming out ceremony. This is far less awesome and affirming than it sounds: back in those days, it just meant going to see the monarch and going to a few balls and, you know, basically telling the world 'I'm here, I'm grown, it would no longer be scandalous to marry me.'

So, in 1833, aged seventeen, Ada was sent to London to meet the Queen and go to a bunch of posh parties with a view to find some dude to marry. Instead, she found some dude who would change her life in a very different way – a forty-one-year-old crank named Charles Babbage.

Now, Charles Babbage was way, way closer to that prototypical nerd character we were talking about earlier. Like, in university he founded a society devoted to reforming the notation for calculus, and was annoyed that his maths degree didn't contain enough maths. He was also, frankly, a massive killjoy who hated street music, kept a running tally of things he'd seen people doing that he considered 'nuisances', and led an unsuccessful crusade to ban hoop-rolling, which you may remember as being the game you see poor kids playing when films are trying to really hammer home how little joy they have in their lives.

But for the young Ada, none of that mattered, because Charles had something far better than a silly hoop and stick game – he had the Difference Engine.

Well, to be more accurate, he had a prototype of the Difference Engine. A decade into its construction, he hadn't been able to finish it yet on account of having neither enough money of his own nor any rich friends who thought a gigantic 4-tonne steam-powered computation engine – a 'computer', you might perhaps call it – was worth investing in.

Ada was different. She immediately saw the possibilities of the machine, calling it 'a gem of all mechanism', and she devoured the notes and diagrams Charles sent her on his invention. And Charles, too, was grateful for the kindred spirit he found in Ada, telling her he could 'Forget this world and all its troubles and if possible its multitudinous Charlatans—every thing in short but the Enchantress of Number', which must have been nice if only for the nickname upgrade from 'it'. And when he got started on his next invention, the Analytical Engine, Ada became nearly as obsessive as he was about it – when tasked with translating an article about the Engine from an

Italian fan into English, for example, she diligently did so … and also added so many notes and suggestions that the finished product was literally four times as long as the original.

By the way, there's one of these additions that deserves special mention. Not for nothing is Ada Lovelace known as the 'mother of computer programming', and it was this translation – specifically, it was in Note G of the translation, because she genuinely added so many that she needed an appendix to order them properly – that established that title for her. Note G contains what is generally accepted to be the first ever program: a detailed algorithm for the calculation of Bernoulli numbers.*

To Charles, the Analytical Engine was basically the Difference Engine on steroids. All the little kinks that his first machine had ended up suffering from would be fixed in the new one; the Analytical Engine would be able to take on any maths problem you cared to set it. A super-calculator, essentially.

But to Ada, the Analytical Engine was a portal to the future.

'[The Analytical Engine] might act upon other things besides number, were objects found whose mutual fundamental relations could be expressed by those of the abstract science of operations, and which should be also susceptible of adaptations to the action of the operating notation and mechanism of the engine,' she wrote, before immediately predicting the invention of the iPod:

'Supposing, for instance, that the fundamental relations of pitched sounds in the science of harmony and of musical composition were susceptible of such expression and adaptations, the engine might compose elaborate and scientific pieces of music of any degree of complexity or extent.'

So there's no doubt that Ada was a visionary. But while she may have been able to foresee things nobody else could imagine when it

---

* These are a sequence of numbers, which if you're a mathematician you'll already have heard of, and, if you aren't a mathematician, you will literally never need to know about, so I'll omit the explanation.

came to computers, there were other things she never figured out – even though everybody knows it. And one of the biggest was this: there is no 'system'.

The first time Ada went to the racetrack, she had been a teenager. It was not long after her first meeting with Babbage, and she was on a trip out with her mother. But, starting in her twenties, after a bout of bad health that her doctor had diagnosed as 'fucked if I know, have some opium', and knowing that the Analytical Engine would never get off the ground without funding, Ada began gambling compulsively.

This would have been less of a problem if she was any good at it, but she wasn't. She was so bad at gambling that she lost thousands and thousands of pounds, sometimes in a single race – she once lost £3,200 in one go, which in modern money is more like £200,000. And, sure, maybe you're thinking 'well, that's just what rich people do', and that's true, but here's the thing: Ada could *absolutely not* afford this. Like so many gambling addicts, she was convinced she had a foolproof system that would let her win big, and also like so many gambling addicts, she ended up having to beg her loved ones – mostly her mum – time and time again to bail her out.

But perhaps the crowning glory of Ada's inability to hold on to money came right at the end of her life, when, essentially on her deathbed, she confided to her mother that she had pawned her husband's family jewels for gambling funds – the ones in their place were paste replicas.* Seeing that Ada was ashamed, and repentant, and in a whole lot of debilitating pain due to late-stage uterine cancer, Annabella dug deep and paid for all the jewels to be replaced before Ada left this mortal plane.

Her debts finally paid off, and mere weeks away from death, Ada then immediately pawned the replacements for gambling money and lost the whole lot on bad bets at the racetrack with a guy she was, let's

---

* A fact which, amazingly, her husband never seems to have noticed.

face it, almost certainly having an affair with.* Her mother was, as you might imagine, not pleased, but there's a limit to how mad you can get at your only child while she's in so much pain you've had to literally pad her bedroom walls with mattresses to stop her injuring herself while spasming in agony, and Annabella once *again* paid for the jewels to be replaced.

This time, Ada was not to be allowed near the jewels, even as sick as she was. But by this point, she had only days left to live: not enough time to stage a jewel heist and win big on the Grand National. Instead, she made a mysterious bombshell confession† to her husband that made him suddenly leave her – and this was a guy who had put up with a *lot*, for the record – and died. She was thirty-six – the same age her father had been at his own death.

Almost immediately, the world pretty much forgot about Ada Lovelace. It wasn't until the late twentieth century that people started taking an interest in her legacy once again – and even now, when she's rightly recognised as a hero of STEM and even has an annual Google Doodle in her honour, she's probably not as well-known or accepted as her contemporaries like Faraday or even Babbage himself.

The obvious reason for that is that she was a woman – and not just a woman, but a swearing, drinking, gambling geek of a woman. She was brilliant, and she was an idiot, and she was probably a bit of a bitch in all honesty, but she had something that very few people have: the ability to not just imagine a better future, but to figure out how to make it happen.

And a shit-ton of paste-replica jewels.

---

* His name was John Crosse, and he was the son of Andrew Crosse, an amateur scientist most famous for thinking he had created life, like some kind of god, but actually having found cheese mites, like some kind of guy who had cheese in his house (it didn't quite go down like that in reality, but that's the story that made him famous). There's little hard evidence that the pair were actually having an affair, but that's largely because Ada told John to burn all the correspondence they'd ever sent each other after she died, which is suspicious to say the least.

† See note above.

15.

Galois Hunting

Have you ever seen the movie *Good Will Hunting*? It's this late-nineties film about a twenty-year-old mathematical genius named Will Hunting – which is lucky, because the title pun wouldn't work otherwise – who suffers from an unfortunate addiction to repeatedly screwing himself over. Over the course of two hours, we see the young Will get drunk, get in fights, be arrested, ruin good relationships, insult the NSA, sabotage lucrative job opportunities, and use matrices to solve problems he was clearly told were advanced Fourier systems.* Eventually, the soothing wisdom of his psychiatrist Robin Williams finally gets through to him, and his story ends with him driving off into the sunset with the love of his life.

Good Will Hunting isn't based on the life of any real mathematician, but change a few details around, maybe *Les Mis* it up a bit, and there is one to whom it gets pretty close: Évariste Galois. Now, if you

---

* Which … aren't a thing, by the way. I know I have an advantage here, having spent an embarrassingly large fraction of my life studying maths in a formal setting, but it's just *weird* that the writers of a movie specifically about maths would slip up this bad. For the record, the problem that Will actually solves is a fairly elementary graph theory question, for which matrices are a perfectly appropriate tool.

look him up online, you'll find there's only about two pictures in existence of Galois, and both look like a pencil drawing of a vaguely annoyed adolescent ghost. You'll also find a lot of adjectives like 'brilliant', 'extraordinary' and 'genius', and Wikipedia articles with titles like 'List of things named after Évariste Galois'.

That's odd though, right? I mean, if he was such an amazing mathematician, shouldn't we have a few more clues as to what he looked like? Y'know, rather than just some etching of a moody kid and a pencil drawing that, now we're reading the caption a little closer, turns out to have been done from memory by somebody who hadn't seen him for sixteen years? Did he just ... stop existing when he reached twenty or something?

Well ... yeah, actually. That's exactly what happened. On 30 May 1832, early in the morning, Évariste Galois was shot in the abdomen in a duel over a girl. The shooter and his seconds fled the scene, leaving him to bleed out alone until a passing farmer found him. He was taken to a hospital, where he died at 10 a.m. the next day. His last known words were to his younger brother: 'Don't cry, Alfred! I need all my courage to die at twenty!'

The death of Galois is one of the most infamous tragedies of the mathematical world – even more so than Pythagoras and that field of beans. Baby mathmos are brought up hearing the legend of his last night, spent frantically writing down every theorem, conjecture and proof he could remember in a small notebook that he bequeathed to a friend.

'Ask Jacobi or Gauss* publicly to give their opinion, not as to the truth, but as to the importance of these theorems,' he requested in a letter accompanying the notes. 'Later there will be, I hope, some people who will find it to their advantage to decipher all this mess.'

---

* If you're not familiar with these names, this is basically the equivalent of your mate Barry who was always pretty good at PE sending you a video of a trick shot he says he invented and asking you to send it over to Ronaldo for comment.

In the years that followed there would indeed be people who found his notes to their advantage – hundreds of them, in fact, if not thousands. After his death, the great mathematician Liouville discovered in those last-night scribblings a proof 'as correct as it is deep' – his words – of an unsolved problem underpinning an area of maths we now know as group theory.* He found important contributions in analysis and number theory, all scribbled down alongside doodles of the name Stéphanie – the girl for whom he was going to his incredibly pointless death – and poignant author's notes, like 'There is something to complete in this ... I do not have the time.'

Group theory is an area of algebra so fundamental to maths that it touches just about every aspect of the modern study, and without Galois, it might not even exist. There's even a whole area of maths named after him: Galois theory. Honestly, there's no telling what great heights this young mathematician could have reached if it hadn't been for that one person who had it out for him his entire life: Évariste Galois.

Galois was born in France in 1811, which meant two things: one, that he was living in a time and place marked by near-constant political chaos, and two, that everybody was both extremely emotional and at least slightly drunk at all times. Both of these facts would play important roles in how his life unfolded.

Even against the backdrop of the time, though, the Galois family stood out as being a bit ... well, *extra*, and a quick rundown of the young Évariste's life contains levels of drama that would make a soap-opera villain take a step back and tell you to calm down. By the time he entered school at age twelve, for instance, he had seen France go through no fewer than four regime changes, some lasting only a

---

* For those nerds in the audience who want to know exactly what the problem was: the question was to decide when a polynomial equation can be solved with radicals – basically: 'which equations are easy to solve?' Honestly, I could go into detail about what exactly the statement means, but the truth is that, if you're a maths junkie, you probably already understand it, and if you're not, then you *truly* will never need to.

matter of months. Then, literally within his first term, there was a student rebellion against the new provisor that led to the expulsion of some forty pupils.

At fourteen, he encountered maths for the first time, and began a relationship that would rival Romeo and Juliet's for its passion and drama. At fifteen, he was reading university-level maths textbooks and original research papers like they were the latest YA novel. He started getting a reputation for being obsessive and confrontational; his school's director of studies basically told his parents to pull him from formal education since 'he is wasting his time here and does nothing but torment his teachers and overwhelm himself with punishments'.

At seventeen, Galois was simultaneously failing literature – his Baccalaureate examiner reported that he 'knows absolutely nothing. I was told that this student has an extraordinary capacity for mathematics. This astonishes me greatly, for, after his examination, I believed him to have but little intelligence' – and getting maths papers published in top-tier academic journals.

And at eighteen, just a few weeks before he was meant to sit his entrance exams for the École Polytechnique, his father suddenly killed himself after a local priest framed him for the crime of writing some mean poems. Galois, unsurprisingly, failed to gain entry to the school. Oh – but don't think his big old maths brain would have sailed him through otherwise; he was notoriously bad at explaining himself to his tutors, and at one point in the examination he actually threw a blackboard eraser at the head of one of the professors, which is one of those things that tends to bias a panel against you.

You see what I mean about Galois being his own worst enemy? Here we have a kid who's a bona fide genius, already published in multiple journals, in correspondence with mathematicians who are already legendary figures in their own right – and apparently doing everything he can to get himself rejected from every part of the academic world as quickly as possible. Did this kid even *like* maths?

The thing was, yes, Galois loved maths – but he loved revolution even more, and he chased it with a fervour that could almost be called suicidal. He had tried twice to get into the École Polytechnique, not because it was the best in Paris for studying maths – it was, but it also had a reputation for radical republicanism among its students. They had famously taken to the streets in protest after Napoleon changed his title from First Consul of the Republic to Emperor, and again a decade later when he had tried to reinstate himself during the Hundred Days after his escape from Elba. And while he couldn't have known it when he was applying for entry, they would also take part in the July Revolution of 1830 – the fifth national upheaval in Galois's short life, which saw King Charles X overthrown and replaced by his cousin, Louis Philippe.*

It must have been irresistible to the young firebrand Évariste. Which might be why he took it so hard when, instead of letting him in without question, the École Polytechnique made him do things like 'explain his logic' and 'stop throwing things at our heads', and then *denied* him a place and forced him to attend the École Normale instead.

There, while the Polytechnique students rioted for democracy in 1830, Galois – along with all the other École Normale students – was locked inside the school by the principal to stop him taking part. Perhaps you're thinking the teacher did him a favour, but Galois wasn't going to let anybody stop him from running headfirst into every brick wall life threw his way. If he couldn't jeopardise his future by running riot in the streets, he was damn well going to do it by badmouthing his anti-protest teacher in the press. And that's the story of how Galois got himself expelled from his back-up school.

So, aged eighteen, without an education, and with no job or father to support him, Galois was set loose upon the world. He decided to try to eke out a living as a maths tutor, but – in case you weren't getting this yet – he didn't exactly have the kind of patient, empa-

---

* Who would go on to be overthrown himself eighteen years later in the February Revolution. I told you this time period was crazy.

thetic personality you need to be a good teacher, and all his students pretty quickly unenrolled themselves from his classes. Things were getting so dire that Sophie Germain, the brilliant number theorist, wrote to a friend at the time that 'they say [Galois] will go completely mad. I fear this is true.'

But you know the old saying: when you're at rock bottom, there's nowhere to go but up.

Évariste, however, did not know that old saying, and he just carried on storming his way downwards. How, you ask? How could he possibly make things worse for himself? By getting drunk and publicly threatening to kill the new king, of course!

It was 9 May 1831, and around 200 republican revolutionaries had met for a gigantic dinner party, which seems less than clandestine but what's the point in being a nineteenth-century French revolutionary if you don't get to enjoy the food? Alexandre Dumas, who you may be aware of as the author who gave us *The Count of Monte Cristo* and *The Three Musketeers*, wrote that 'it would be difficult to find in all Paris two hundred persons more hostile to the government than those to be found reunited at five o'clock in the afternoon in the long hall on the ground floor above the garden', so of *course* Galois took the opportunity to show his face in the crowd. Not just in the crowd, in fact: he stood up, raised a glass in one hand and a knife in the other, and 'toasted' the King to the raucous cheers of the assembled republicans.

Now, in *Good Will Hunting*, there's a scene where Will ends up in front of a judge too, and despite arguing fairly convincingly in his own defence, he – spoiler! – ends up being sentenced to jail time. Well, Galois did the same – except instead of protesting his innocence with sophistry and legal precedents, when Galois was asked by the judge, 'Did you do it?', he answered essentially: 'Yes, and I'd do it again.'

Despite nine out of ten lawyers agreeing that this is probably not the best legal defence, Galois was once again thrown a lifeline: the jury, apparently swayed by the fact that he was still just a kid, acquitted him. Which he decided to celebrate by, less than a month later, dressing up in the uniform of an illegal anti-monarchist battalion, arming himself with a rifle, dagger and several pistols, and joining a Bastille Day demonstration.

Of course, today, taking part in Bastille Day celebrations would be seen as an act of patriotism; it is, after all, France's National Day. But, for most of the nineteenth century, the memorialising of a day intimately associated with the overthrow of the monarchy was – kind of understandably, from the current king's point of view – considered sedition. As a vague comparison, you could think of how people might have reacted to somebody deciding to put on a red coat and wander round 1780s New York armed to the teeth and waving a Union Jack.

This time, his trial didn't go so well. He spent nine of his last twelve months in prison, during which time he allowed his fellow inmates to goad him into getting drunk, mournfully prophesised his own pointless death with suspicious accuracy* and occasionally tried to thwart said predictions by killing himself instead.

His genius – and his ability to make enemies in high places – clearly wasn't blunted by his lodgings, though, because it was also in prison that he wrote one of the most remarkable documents in the history of mathematics: his *Préface*. It's a five-page, well, preface, to an independently published paper titled 'Two Memoirs in Pure Analysis', which doesn't sound particularly controversial, but in Galois's hands was practically a declaration of war against establishment mathematics.

---

* 'And I tell you, I will die in a duel on the occasion of some *coquette de bas étage*. Why? Because she will invite me to avenge her honour which another has compromised,' he reportedly told his co-confinee and future statesman François-Vincent Raspail. It's hard not to feel sorry for the poor lad at this point. He continued: 'Do you know what I lack, my friend? I confide it only to you: it is someone whom I can love and love only in spirit. I have lost my father and no one has ever replaced him, do you hear me … ?'

I say to no one that I owe to his counsel or to his encouragement all that is good in my work I do not say it because it would be a lie. If I were to have to address something to the leading men of the world or to the leading men of science (and at the present time the distinction between these two classes of people is imperceptible), I swear that it would certainly not be thanks.

I owe it to the one group to have caused the first of the two memoirs to appear so late, to the other to have written it all in prison. Why and how I am kept in prison is not my subject, a stay that one would be wrong to consider as a place of contemplation, and where I have often found myself astonished by my carefreeness at not closing my mouth to my stupid *Zoïles* [spiteful critics]; and I believe I can use this word *Zoïle* without compromising my humility, so low are my adversaries to my mind.

I repeat: five pages of this. And why? Because he had a paper rejected – an experience I can assure you not one professional mathematician in the world has not faced.

'The first memoir is not new to the eye of the master,' reads the *Préface*. 'An extract sent in 1831 to the Academy of Science was submitted for refereeing to M Poisson, who just said at a meeting that he had not understood it. Which to my eyes, fascinated by the author's self-confidence, proves simply that M Poisson did not want or was not able to understand it; but certainly proves in the eyes of the public that my book means nothing.'

Poisson, one of the most famous mathematicians ever, had done him dirty, he claimed. It wasn't that he was bad at explaining himself or too combative or anything – it was simply that every mathematician and mathematical body in the country had it out for him, so why even bother.

He went on: 'In the scientific world the work that I am submitting to the public would be received with a smile of compassion ... the most indulgent would accuse me of awkwardness; and ... after some time I would be compared with Wronski or these indefatigable men who every year find a new solution to the squaring of the circle.' (Apparently he had forgotten about not compromising his humility.) Then the deeper scars came out. Now in full rant mode, he started on 'the laughter of the examiners of candidates for the *École Polytechnique* (who, by the way, I am astonished not to see, each one of them, occupying an armchair in the Academy of Science, for they certainly have no place in posterity)'; they would not understand how this 'young man twice failed by them might also have the pretension to write ... books of doctrine.'

And just to make sure you really understand how extremely Not Mad he is, he assures the reader that 'it is knowingly that I expose myself to the laughter of fools'.

And then, Fate intervened. And Fate looked like a cholera outbreak.

Don't worry: Galois didn't get it himself – his story isn't quite *that* bleak. In fact, this is where things started looking up for him for a little – very little – while. He had just one month left on his sentence, and he had been moved, along with all the prisoners, to a guesthouse outside of the city to avoid the epidemic. And with the guesthouse came – Stéphanie.

What happened next is probably familiar to all of us – well, apart from the ending. Galois thought he was in love with Stéphanie; Stéphanie did not reciprocate. She tried to let him down gently, but he didn't take a hint, and eventually her fiancé made himself known. But whereas these days the worst we'd expect from this set-up is a punch-up outside a pub, 1830s France took a much more 'pistols at dawn' approach to settling disagreements.

Galois died on 31 May 1832, but the exact details, including even the identity of his shooter, remain shrouded in mystery. Because of this, you can still find people today who think he was the victim of some elaborate conspiracy involving the royal secret police – they paid this Stéphanie to lead him on, they suggest, then shot him in the resulting duel because of his republican leanings.

But the truth is that there was no conspiracy. Who would need one? Galois died exactly as he lived: taking every bad decision he saw, making as many enemies as possible, and being *intensely* overdramatic. He was a genius at mathematics, but when it came to living life he was dumb as a box of rocks, and it got him killed. That's all.

If only he had gone to see Robin Williams. What a waste.

16.

John Couch Adams Ignores His Mail, Loses Neptune

Here in the scientifically enlightened twenty-first century, there's no longer any doubt as to how many planets are out there rolling around in our solar system. It is, of course, eight – nine, if you're a Pluto truther, anywhere between thirteen and a couple of hundred if you count dwarf planets, or around 800,000 if you want to be a real dick about it and include minor planets as well. Easy.

Two hundred years ago, however, things were simpler: there were seven planets. And while astronomers of the day knew there was *something* wrong with their model, the problem was at least one that was fun to say: Uranus was wonky. The newly discovered farthest-most planet just wouldn't behave itself, wandering about like a drunken space hopper and turning up in places where it frankly had no business to be. Every so often some scientist would declare they'd discovered the source of the issue* and everything would be fixed again – only for the whole sorry mess to play out afresh a couple

---

* Invariably, this was revealed to be 'all scientists except for me are idiots'.

of years later when Uranus decided to go for another unscheduled romp through the cosmos.

After a few decades of systematically ruling out every possible solution for the cosmic conundrum, the world was left with two options: either all science ever invented was wrong, or there was some mysterious, as-yet-undiscovered planet, even further out, whose gravitational force was somehow powerful enough to have been making Uranus walk funny for years.

Figuring that they really didn't have the time to re-invent thousands of years' worth of scientific enquiry, a few optimistic stargazers instead decided to begin the hunt for a new celestial body that could finally explain Uranus's sloppy appearances.

One of these optimists was a young mathematician named John Couch Adams. Now, whatever image you have in your mind of a 'genius', Adams was it. He was mostly self-taught in maths – pretty much his only option, given the educational opportunities available to a farm kid in Georgian Cornwall – and was by all accounts a genuine prodigy. As a boy, his father would take him out in public and show off his abilities, even pitting him against other children and adults, presumably in a bid to win his son the coveted 'least popular kid in town' award. From Cornwall, he went on to study at Cambridge as a sizar – Oxbridge for 'poor' – eventually graduating with a grade in mathematics reportedly twice as high as the guy who came second. For this achievement, he was granted the title of Senior Wrangler, which must have made all that hard graft finally worth it.

Now, like all nerds, Adams owed much of his academic success to a painstaking avoidance of anything cool, and also like all nerds, his idea of what constituted 'cool' was probably not what you might expect. You see, while many students might attribute their good grades to a life of celibacy and sobriety, Adams's mathematical success at university was – at least according to his own reports – thanks to his abstinence from astronomy.

Which meant that, when he finally graduated, he went wild, mainlining the stuff so hard that had it been something other than a relatively abstract field of scientific study he probably would have ended up in rehab.

'[I have] formed a design,' he wrote in July 1841, 'of investigating, as soon as possible after taking my degree, the irregularities in the motion of Uranus ... to find whether they may be attributed to the action of an undiscovered planet beyond it; and if possible then to determine approximately the elements of its orbit ... which would probably lead to its discovery.'

It's important to recognise just how ambitious this plan really was. Every planet so far known to science had been found, basically, by just looking upwards and pointing them out – in fact most, easily seen by the naked eye, had been known for millennia. How they interacted with each other, the routes they took across the sky – these were all worked out through years of careful observation. Even Uranus, the maverick recluse of the bunch, owed its recent discovery to a guy and his telescope.*

What Adams was proposing, however, was radically different. He hadn't found a new planet – he didn't even have any real proof that one existed. All he had was a badly-behaved Uranus, and the hope

---

* The guy was William Herschel, who found it while searching the skies from his garden in Bath, and he actually thought he had found a comet. It should be noted that Uranus had been observed many times before this – possibly as early as the second century BCE – but thanks to its slow orbit and dimness, it had generally been assumed to be a star. However, thanks to Herschel's observations, astronomers realised for the first time that the faint light in the sky they had been ignoring for centuries was in fact a seventh planet, and, in honour of this, he was asked by the Astronomer Royal to bestow the new world with a name.

Unfortunately, the name he chose was George. Everybody quite quickly decided that actually perhaps discovering a planet *didn't* necessarily mean you deserved to name it, and started coming up with alternatives. *Uranus* was one of the first suggested – named after the Greek god of the sky – and over the next seventy years Herschel's beloved *Georgium Sidus* was slowly but surely renamed with the moniker we all know and love today.

that some massive, far-off astronomical body that had inexplicably managed to remain unseen to human eyes throughout thousands of years of astronomical observations was somehow responsible. The idea that someone could work backwards like this – that you could start with the observed effect on one planet's behaviour and use it to figure out the size and orbit of the neighbour causing it – was something that had simply never been done before.

It was also, even for someone like Adams, an incredibly complicated problem. It was all well and good assuming that the weirder aspects of Uranus's orbit were caused by a rogue planet, but this also meant that all the supposedly good information collected over the years couldn't really be trusted either. After all, everything scientists thought they knew about Uranus had been based on the idea that it was the furthest thing out in the solar system. If Adams was going to assume that was wrong, it meant starting again firmly from square one – not just with the new mystery planet itself, but with the very information he wanted to use to discover it.

And yet, amazingly, within just a couple of years, Adams had a solution: a set of celestial coordinates that should contain the elusive eighth planet. What he didn't have, however, was a telescope. Confident of his results, he reached out to the highest astronomical authority in the country for help – and that's where everything started going wrong.

For the last decade, the post of Astronomer Royal, director of the Royal Observatory in Greenwich, had been held by a man named George Biddell Airy. Now, there was no doubt that Airy was a smart guy: just like Adams, he came from a poor farming family and attended Cambridge as a sizar, and just like Adams, he graduated top of his class in maths, winning the title of Senior Wrangler and the university's coveted Smith's prize a full twenty years before Adams would do the exact same thing. He was appointed Lucasian Professor of Mathematics – a position later held by Stephen Hawking – just three years after graduating; two years after that he was made

Plumian Professor of Astronomy and director of the newly established Cambridge Observatory. And at the Greenwich Observatory, his directorship was marked by a period of reorganisation and renovation, much of which was thanks to his own engineering prowess.

On the face of it, he was the perfect person for Adams to have written to. But, unfortunately for both of them, Airy had one major flaw that would end up losing them everything: his terrible, terrible personality.

Airy was a snob. Airy had always been a snob. He was an unpopular teacher's pet at school, and his own biggest fan at university, being known for what biographers have diplomatically called an 'immodestly high' opinion of himself. He repeatedly proved almost incapable of achieving even minor career progression without starting at least one life-long rivalry with someone or other, and as Astronomer Royal he was rude, overbearing, and so hostile to independent thought that a grand total of zero young scientists were trained at Greenwich during his tenure.

And one thing that Airy simply did not truck with was abstract flights of fancy like 'Hey, maybe there's a mysterious eighth planet out there screwing things up, but we just can't see it.' He had actually been one of the very first people to hear of the hypothesis: a decade earlier, when the Reverend (and amateur astronomer) Thomas John Hussey originally came up with the idea, it was Airy that he turned to for help working out the fine details.

'Your opinion may determine mine,' he wrote in an 1834 letter to the already eminent Airy. 'Having taken great pains last year with some observations of Uranus, I was led to examine closely Bouvard's tables of that planet. The apparently inexplicable discrepancies between the ancient and modern observations suggested to me the possibility of some disturbing body beyond Uranus, not taken into account because unknown … What is your opinion on the subject? … If the whole of this matter do not appear to you a chimaera … I shall be very glad of any sort of hint respecting it.'

Unfortunately for Hussey, however, Airy was exactly that kind of awful person who takes a request for advice as some kind of invitation to actually give it.

'It is a puzzling subject,' he replied, '... not yet in such a state as to give the smallest hope of making out the nature of any external action on the planet ... [I doubt] any extraneous action.'

It turned out, of course, that Airy, evidently yet unsurprisingly a devotee of the old 'everyone sucks but me' school of scientific enquiry, had a better explanation for the problems with Uranus: specifically, that every astronomer and mathematician involved in the study of the planet so far had simply managed to screw up somewhere along the way. It was just lucky for them that he knew a guy who might be willing to help them out. Y'know, if he got a chance. He'd probably already sorted it out, actually. You might know him. Name's George Airy.

'The state of things is this,' he wrote. 'The observations would be well reconciled if we could from theory bring in two terms; one a small error in [Bouvard's observations], the other a term depending on twice the longitude. The former, of course, we could do; of the latter there are two ... the first I have verified completely (formula and numbers); the second I have verified generally, but not completely: I shall, when I have an opportunity, look at it thoroughly.'

And then, just in case Hussey's dreams hadn't been *completely* demolished yet, he delivered the final blow.

'If it were certain that there were any extraneous action,' he wrote, 'I doubt much the possibility of determining the place of a planet which produced it. I am sure it could not be done till the nature of the irregularity was well determined from several successive revolutions.'

In other words – since a single revolution of Uranus takes just over eighty-four years – even if Hussey was right, there was no point pursuing it. It would take hundreds of years to be able to prove.

And if Airy wasn't going to accept an eighth planet from the dear old Reverend Hussey, the respected gentleman scholar and friend of such luminaries as Darwin and De Morgan, he certainly wasn't going to accept it from some twenty-six-year-old nobody fresh out of a maths degree. So, between September and October 1845, and despite a glowing letter of introduction from James Challis, his successor at Cambridge, Airy managed to miss no fewer than three potential meetings with Adams to discuss his results.

Now, the first time Adams dropped by at Greenwich, Airy had a pretty watertight excuse for not showing up: he was in France. The second time, though, he had less of a leg to stand on. He was indeed in London, but he was out, busy stoking one of his vintage Science Rivalries – this time with Charles Babbage.[*] Adams left his card, and a message saying that he would call back in an hour.

As promised, Adams came back once again, and once again was told that Airy couldn't see him. Yes, he was in, and, no, he wasn't particularly busy – but he was eating dinner, and wasn't going to let his pie and kidneys[†] go cold just for the sake of the biggest potential breakthrough in astronomy for the last sixty years. Finally taking the now unmissable hint, Adams decided to leave for good, but not before leaving a three-page note detailing his findings so that Airy could, should the mood strike him, hunt down the new planet on his own.

A couple of weeks later, Adams received a letter. Airy had finally read his note and grudgingly admitted that his predictions were 'extremely satisfactory' – but he needed more before he'd be properly convinced. 'I would be very glad to know whether this assumed perturbation will explain the error of the radius vector of Uranus,' Airy had written, referring to the distance between the wandering planet and the Sun. 'This error is now very considerable.'

---

[*] OK, it's like this: Airy and Babbage had been nemeses ever since 1826, when they were both up for the position of Lucasian Professor of Mathematics. After a tense election in which Babbage even threatened, for reasons lost to the mists of time, to bring legal action against his opponents, Airy proved successful, and a rivalry was born.

[†] The record doesn't actually say what he was eating, but he was an old Victorian dude, so probably that.

It's strange how sometimes the most innocuous things can set off a fight, isn't it? Airy's question was a perfectly reasonable one: would Adams's proposed new planet explain why Uranus wasn't where it was supposed to be? But to Adams, it was the last straw – almost an insult of a question. *Of course* the existence of Neptune would affect the orbit of Uranus, that was literally his starting assumption! Those predictions for the radius vector that Airy was so concerned with were based on calculations that never accounted for Neptune, weren't they? So, yes, Airy, *obviously* they will be wrong. As far as he was concerned, this was a trivial, stupid question. He never replied.

The trouble was that, to Airy, still the sceptical traditionalist, it was nothing less than *the* question – the point of the whole investigation. And from his perspective, he was simply following the scientific method – a process which pretty famously does *not* start with 'assume my conclusion is correct': unlike Adams, Airy hadn't started out convinced that Neptune existed, and so he wanted a damn good reason to accept it before he ran around claiming strange things about invisible planets messing with Uranus.

Really, all Adams had to do was run the numbers again and see what happened to Uranus's orbit. But he didn't. And so the story ended there, with Adams's historic discovery, and any evidence of an eighth planet, doomed to gather dust in a desk drawer in Greenwich for all time.

Or, at least, that's what could have happened. But in an incredible coincidence, at exactly the same time that Airy was brushing Adams off with ever-weaker excuses, an astronomer and mathematician by the extraordinarily French name of Urbain Jean Joseph Le Verrier had been working on the very same problem that Adams had just solved.

In the same week that Adams decided to give up on the Astronomer Royal, Le Verrier was presenting a memoir to the Parisian Académie des Sciences arguing that current scientific theories could not explain the strange behaviour of Uranus. Six months later, in June 1846, he followed this with another, in which he explained not only that the answer was an eighth planet, but also where to find this mysterious 'Planet X'.

His results were less than a single degree out from Adams's.

There were now two sets of near identical estimates for the position of Neptune: one by Adams, one by Le Verrier. But while the results were the same, there were some crucial differences between the two mathematicians. Le Verrier was older than Adams, and had a respectable set of serious publications behind him, while Adams was still a virtually unknown graduate student. For scientific snob Airy – the only person who knew about both sets of results, remember – it was obvious whose discovery this new planet really was – and it wasn't the twenty-six year old upstart who couldn't even be bothered to reply to a letter.

Now, in case you're reading this and thinking 'Well, this is some bullshit right here, Adams was robbed!', it's worth pointing something out: Le Verrier didn't have it *entirely* easy. Just like Adams, he was ignored when he first wrote to the Paris Observatory asking them to look for his planet. But Le Verrier didn't just have age and experience on his side. He had gumption too. And when he was rejected by Paris, he didn't just pout and go back to his studies – he went straight to the Berlin Observatory instead, where an assistant by the name of Johann Gottfried Galle found Le Verrier's predicted planet within just a single hour of searching. Le Verrier was awarded the Copely medal by the Royal Society and – probably even more impressive on the old CV, this one – the new celestial discovery was named Le Verrier in his honour. The eighth planet had been found, and it was inescapably French.

However, if you know anything about the solar system (or if you don't, but have been paying the bare minimum amount of attention throughout this chapter), you'll know that we do not, as a general rule, talk about the planet Le Verrier. So how did the new discovery come to be known as Neptune? Well, just like a surprising number of other world-changing scientific discoveries, it's all thanks to a

John Couch Adams Ignores His Mail, Loses Neptune • 163

frankly incredible amount of passive-aggressive pettiness between various scientists.

Considering he'd just had the biggest discovery in sixty years stolen from right under his nose, Adams proved remarkably sanguine about Le Verrier's success, taking the view that, well, at least now the eighth planet was out there, so to speak, and that meant lots of exciting new research possibilities for him to get his teeth into. But by this point, word had got out about his earlier calculations, and – even though Airy still had no intention of speaking up for the young upstart who interrupted his dinner – there was one scientist back in England who felt Adams deserved to be, at the very least, a footnote in the history books. Seeing news of Le Verrier's discovery in *The Times*, Sir John Herschel,* one of the country's foremost astronomers, dashed off a letter to the British magazine *The Athenaeum* saying how brilliant Le Verrier's work was and congratulations to him but, by the way, and, no offence, we sort of knew about this for a while already, actually.

Le Verrier was livid.

'When he scrupled not to put into print that my calculations were not sufficient,' he wrote, in an angrily poetic letter published in the *London Guardian*, 'did he not perceive that he was bringing discredit on his own scientific penetration† when he attacked a calculation which observation had confirmed in so signal a manner?'

Of course, Herschel had done no such thing, and he quickly wrote back to the *London Guardian*, assuring readers that 'the prize is [Le Verrier's] by all the rules of fair adjudication, and there is not a man in England who would grudge him its possession' – but it was too late. The French press were furious. Accusations flew, asking why there had been no mention of Adams's work before now, and who was this Adams fellow anyway, and wasn't this all just a transparent attempt by Herschel to steal Le Verrier's glory for perfidious Albion?

---

* If the name sounds familiar, it's because his dad was the guy who discovered Uranus in his garden back on page 156.
† Don't blame me, it was apparently just a smutty time in scientific history.

'Wrote to the editor of the *Guardian* in reply to M. Le Verrier's savage letter,' Herschel wrote in his diary for the day. 'These Frenchmen fly at one like wildcats.'

Ultimately, though, the war over Neptune was short-lived, thanks in a large part to Le Verrier himself. See, people will accept some level of dickishness from you if you're demonstrably brilliant, but only up to a point – and Le Verrier, well, he misjudged his allowance. He was indignant, and arrogant, and his insistence, for instance, that the new planet *must* be named after him and all dissent was just jealous rivals trying to steal his discovery did nothing to endear his scientific peers to his cause. Meanwhile, Adams, riding a wave of respect for having predicted a whole entire planet, and handling the controversy over the discovery with the grace of a well-oiled PR company, started to be viewed as a genius who had been unfairly snubbed by incompetent establishment figures like Airy.*

Eventually a consensus was reached† that Le Verrier may well be right, but he was also annoying, and the planet would be named Neptune, and that nice Adams fella should get some recognition for its discovery. And so the epic saga of the discovery and provenance of the eighth planet in the solar system came to a conclusion, and nobody ever fought over it again.

Just kidding! As late as 2004, new articles were being published with titles like 'The Case of the Pilfered Planet: Did the British Steal Neptune?' And you thought the arguments about Pluto were esoteric and pointless.

---

* Nice Guy Herschel, it should be noted, was quietly very upset by Britain's loss of Neptune to the French, but after having 'said enough about it to get heartily abused in France', he admitted, 'I don't want to get hated in England for saying more.'

† As you might expect, however, it didn't quite reach as far as France.

17.

You Really Wouldn't Want to Hang Out with Karl Marx

When you look at that famous photo of Karl Marx – you know the one – it's hard to imagine him as anything other than a dour old man, ready to scold you for eating more than your assigned share of the communal peanut butter. You know, the kind of guy who corners you at a party and subjects you to a long, dry lecture about economic theory when all you asked was how he knew the host. And he's drinking water.

Well, while it's true that you probably wouldn't want to hang out with him, the real reason is because a night out with Marx could carry a genuine risk of ending in a police chase. And that's not because of his politics or radical activities or anything – or at least, not directly. It's because he was, basically, a bit of a lager lout.

It's a piece of political history that's become infamous: the night Karl Marx got drunk in central London and smashed up a bunch of streetlamps. And like so many stories that end in the nineteenth-century equivalent of 'so we legged it and that's how we avoided criminal damage charges', it happened on a pub crawl with a couple of mates from uni.

'One evening, Edgar Bauer, acquainted with Marx from their Berlin time and then not yet his personal enemy* ... had come to town from his hermitage in Highgate for the purpose of "making a beer trip"', recalled Wilhelm Liebknecht† some forty years after the fateful night.

'The problem was to "take something" in every saloon between Oxford Street and Hampstead Road – making the something a very difficult task, even by confining yourself to a minimum, considering the enormous number of saloons in that part of the city. But we went to work undaunted and managed to reach the end of Tottenham Court Road without accident.'

Now, under normal circumstances, walking from Oxford Street to Hampstead Road – in other words, the length of Tottenham Court Road – should take about fifteen minutes.

A Victorian pub crawl is not normal circumstances. Calling the night's undertaking 'a very difficult task' was, if anything, an understatement from Liebknecht, who failed to specify that this short walk passed a grand total of *eighteen* pubs. Suddenly, reaching the end of the road 'without accident' doesn't seem such a small boast.

---

* Edgar Bauer was a German political philosopher who hung out with Marx while they were at the University of Berlin. Along with other famous names, like Engels, Feuerbach and Edgar's older brother Bruno, they were part of a club called the Young Hegelians on account of their being a) young and b) dedicated to the philosopher Hegel. Despite Bauer starting life as a hardcore anarchist student writing theoretical justifications of political terrorism, in the second half of his life he had transformed into more of a hardcore conservative civil servant, writing theoretical justifications for being a narc. This change of attitude is, unsurprisingly, what caused that enmity between the two philosophers that Liebknecht referenced.

It should be noted that even when Bauer was spying on Marx for the police (that narc reference wasn't an idle insult) and Marx thought of Bauer as some Tory loser, they were still happy to get wasted together – although Bauer did punch Marx right in the face on at least one occasion.

† Liebknecht was a socialist and journalist, and one of the main founders of the Social Democratic Party of Germany (SPD). He may be the least recognisable name of the three for most people, but he may have had the strongest lasting impact on politics: the current Chancellor of Germany, Olaf Scholz, is a member of the SPD.

But there's a limit to how long you can expect three professional dissentients to go without getting into a fight, and that limit is apparently the time it takes to down eighteen pints in various establishments in the West End. When the gang reached the last pub, they found it rather loudly occupied by a group of local Odd Fellows* – apparently quite soused themselves – and they decided to make friends.

'[T]hey at once invited us "foreigners" with truly English hospitality to go with them into one of the rooms,' Liebknecht wrote. 'We followed them in the best of spirits, and the conversation naturally turned to politics – we had been easily recognised as Germany fugitives; and the Englishmen, good old-fashioned people, who wanted to amuse us a little, considered it their duty to revile thoroughly the German princes and the Russian nobles.'

I know, it's hard to believe: a bunch of drunk English blokes showing their arses in a slightly racist manner – though they should have gone after Prussian nobles, from Germany, not Russian ones, if they really wanted to annoy the trio. But Marx and his boys took it in fairly good humour, possibly because the jibes came with unlimited free drinks – until, eventually, some unknown comment caught Bauer the wrong way.

'Edgar Bauer, hurt by some chance remark, turned the tables and ridiculed the English snobs,' recounted Liebknecht. Suddenly, 'Marx launched an enthusiastic eulogy on German science and music – no other country, he said, would have been capable of producing such masters of music as Beethoven, Mozart, Handel† and Haydn.'

---

* Like Freemasons, but with more alcohol.
† Not to imply that Karl Marx ever said anything controversial, but, arguably, the Brits get to claim Handel. Born in Germany, yes, but he moved to Britain aged twenty-seven and spent nearly twice as long again living in London – right next door to Jimi Hendrix, as it happens, albeit separated by a few hundred years. And if further proof were needed as to Handel's nationality, suffice it to say that there exists an Act of Parliament called Handel's Naturalisation Act 1727, which is literally a piece of legislation that was passed in the House of Commons to make Handel British.

Never one to read a room, Marx held forth on the superiority of the Germans over the British – with additional input from Liebknecht and Bauer to help him along – right up until fists started being brandished. And at that point, rather sensibly – and, Liebknecht noted, 'not wholly without difficulty' – the gang cheesed it.

And that's when things got lairy. As Liebknecht wrote:

> Now we had enough of our 'beer trip' for the time being, and in order to cool our heated blood, we started on a double-quick march, until Edgar Bauer stumbled over some paving stones. 'Hurrah, an idea!' And in memory of mad student pranks he picked up a stone, and Clash! Clatter! a gas lantern went flying into splinters. Nonsense is contagious – Marx and I did not stay behind, and we broke four or five streetlamps – it was, perhaps, 2 o'clock in the morning and the streets were deserted in consequence. But the noise nevertheless attracted the attention of a policeman who with quick resolution gave the signal to his colleagues on the same beat. And immediately countersignals were given. The position became critical.
>
> Happily we took in the situation at a glance; and happily we knew the locality. We raced ahead, three or four policemen some distance behind us. Marx showed an activity that I should not have attributed to him. And after the wild chase had lasted some minutes, we succeeded in turning into a side street and there running through an alley – a back yard between two streets – whence we came behind the policemen who lost the trail. Now we were safe. They did not have our description and we arrived at our homes without further adventures.

If you're surprised to find out that the dude responsible for political prose like 'The history of all hitherto existing society is the history of class struggles' or 'All that is solid melts into air, all that is holy is profaned, and man is at last compelled to face with sober senses, his real conditions of life, and his relations with his kind' was actually just a common yobbo, then ... well, I can only assume you don't know much about Karl Marx. See, even as a student, the young Karl was a bit of a tearaway, which is one reason why it took him a total of three different universities to get his doctorate.

The first establishment to welcome – and subsequently evict – Marx was the University of Bonn. He was seventeen, and, as seventeen-year-olds are wont to be, naive as all heck: his plan was to study philosophy and literature, become a poet, and marry a girl called Jenny* he knew from back home. Unfortunately, though, as far as Marx's dad was concerned, being a poet was not a viable career option,† and if Karl wanted to go to university, he would have to switch to law.

---

* This is Karl's chapter, so it wouldn't really be right to go too deeply into the story of Jenny von Westphalen, Marx's wife, but she was quite the character in her own right. Born the daughter of a baron, she ditched a well-to-do lieutenant's proposal to instead marry an upstart socialist revolutionary who was four years her junior, a commoner, and, perhaps most controversially of all, vaguely Jewish.

She was free-thinking, well-educated, and passionate throughout her life for socialism, women's rights and providing for refugees. Her story is quite a sad one – filled with loss and heartache – and she probably would have been a lot better off if she had married the lieutenant. But, without her, none of Marx's legacy would have been possible – she was without a doubt his most dedicated supporter in pretty much every sense.

† Marx Senior was not what you might expect given his son's legacy. First of all, he was a pretty wealthy and well-respected lawyer, who, on one occasion, was recorded as raising 'a toast to the King of Prussia, in the fullness of his omnipotence'. You may think that sounds like sarcasm – and you'd be in good company if you did, since the authorities at the time saw in this glowing praise cause to place him under police surveillance as a potential troublemaker – but he was totally serious.

And he certainly didn't share Karl's passionate idealism. At thirty-five, he undertook a wholesale identity redesign when he changed his name from Hirschel HaLevi to Heinrich Marx and converted from Judaism to Evangelical Christianity – a tactical decision, since Jews in Prussia were effectively banned from holding public office or engaging in professional work. Not that the latter change would have made much practical difference: he was by all accounts neither a religious Jew nor a religious Christian – but it's certainly not one we can imagine Karl 'literally banned from multiple countries because he won't shut up about his principles' Marx making.

No problem, thought Marx, and off he went to Bonn, where he immediately set about taking up those most hallowed traditions of college life: getting drunk and skipping lectures. He joined the Poetry Society and the Tavern Society, grew his hair and beard, and started smoking, and sent a stream of letters home begging for money for undisclosed reasons. And I know none of that sounds too bad on its own, but he also repeatedly fell foul of the local police: at one point he was kept for twenty-four hours in the cells for being drunk and disorderly, his only succour being that the police were apparently fine with his friends sticking around and bringing him more booze and playing cards while he was incarcerated.

He also couldn't seem to stop challenging people to fights. 'Is duelling then so closely interwoven with philosophy?' reads one exasperated letter from Marx Senior after his son was formally cautioned for brandishing a pistol in public. 'Do not let this inclination, and if not inclination, this craze, take root. You could in the end deprive yourself and your parents of the finest hopes that life offers.'

Things came to a head when a drunken sabre duel left him with a cut above his left eye, and Ma and Pa Marx decided enough was enough. 'I not only grant my son Karl Marx permission,' his father wrote in a transfer declaration at the end of his first year, 'but it is my will that he should enter the University of Berlin next term for the purpose of continuing there his studies of Law.'

Now, the University of Berlin was known as something of a swot school, but if the elder Marx was hoping the new setting would rub off on his son, he would be sadly disappointed. It was there that Karl met the Young Hegelians, including Bauer (of later Camden streetlight-smashing fame), and the next few years saw the pair scandalising their classmates with stunts like galloping around the town on donkeys, laughing in church – they were both avowed atheists – and, of course, getting uproariously drunk. Honestly, Marx couldn't even behave himself without ruffling some feathers:

his doctoral thesis, in which he argued basically that philosophy was better and smarter than religion, was too controversial for the professors at Berlin, and he had to submit it to the University of Jena instead.

Now finally graduated, the new Dr Karl Marx obviously cleaned up his act and decided not to stir up any more trouble. Just kidding! Within two years, Karl and Jenny had had to move to a commune in Paris due to Karl's inability to not openly attack the aristocracy in print – and it was here that Marx's lifelong friendship and collaboration with Friedrich Engels had its now legendary beginning: a ten-day non-stop red-wine-fuelled radical economic philosophy debate binge.

Much like his parents and poor Jenny, Engels got the short straw in his relationship with Marx. He was pretty much the main breadwinner for the Marx family, which is impressive considering he wasn't actually part of it – Marx even used Engels's wife's death as an opportunity to beg money off his friend. But that was nothing compared to Marx's biggest request: baby Freddy.

Baby Freddy, or Henry Frederick Demuth, was the son of Helene Demuth, the Marxes' loyal maid and Jenny's closest friend since childhood. He was born in 1851, nine months after Jenny had been away in Holland, leaving her drunk, womanizer husband alone with her friend. The father's name on the birth certificate was left blank, but I think we can all guess who it was.*

This was the Victorian era, though, and there was a certain protocol to these things, so Freddy was duly packed up and sent to live with some other family in the East End. Everyone in the Marxes' lives was told Freddy was Engels's kid – I told you he was short-changed in the friendship – and his reputation suffered for it: Eleanor Marx,

---

* Kind of heartbreakingly, Jenny wrote a letter home at the time telling Karl, 'Oh, if you knew how much I am longing for you and the little ones. I know that you and Lenchen [Helene] will take care of them. Without Lenchen, I would not have peace of mind here.'

one of the seven younger Marxes,* was horrified that the man she had thought of as her 'second father' could so coldly reject a son.

'It may be that I am very "sentimental",' she wrote in 1892, long after the original scandal, 'but I can't help feeling that Freddy has had a great injustice all through his life.'

She was right, of course, but it wasn't the injustice she thought – and a few years later, she would learn the truth. On his deathbed, Engels finally told her the real story of Freddy's paternity.

'[Engels] gave us permission to make use of the information only if he should be accused of treating Freddy shabbily,' recalled Engels's last housekeeper and companion, Louise Freyberger, a few years after the confession. 'He said he would not want his name slandered especially as it could no longer do anyone any good.'

It's hard to know what Marx did to deserve such loyalty. He was by all accounts a pretty shitty friend and husband, and while he was loving and gentle with his daughters, it's hard to think of a man who ditched his illegitimate son in Hackney as a good father. His work ethic was bad in the weirdest possible way: he was somehow both so obsessive that he almost completely ruined his family, and also so nonchalant that when the publishers of *Das Kapital* contacted him six months after his original deadline, he replied that he 'shall have finished soon, having finally begun the actual writing'. As for his manners, he only avoided multiple restraining orders throughout his life by virtue of their not having been invented yet.

People often wonder what Marx would make of his reputation today. They cite quotes like 'all I know is that I'm no Marxist' – something that, according to Engels, he really did say – as proof that he

---

* Karl and Jenny had seven children together, but only three made it to adulthood (and of those three, two killed themselves). The high death toll – even for the time – was likely in a large part due to the crushing poverty the family was forced to live in: it was at times so bad that Marx couldn't leave the house because his clothes had all been pawned, the children were trained in bailiff evasion, and Jenny once had to beg for money for a coffin to bury her baby daughter Franziska.

would have been shocked by the movements his philosophies led to. But the reality is, he probably wouldn't care what we thought of him either way, as long as there was another beer incoming and a few streetlamps to break.

Socialist he may have been, but social he was not. As he once admitted in a game of 'Confessions' with his daughters: '[My] hero: Spartacus; chief characteristic: singleness of purpose; idea of happiness: to fight.'

18.

Charles Darwin: Glutton; Worm Dad; Murderer?

I'm going to tell you a story. It's about a man named Robert FitzRoy, which might surprise you, given the title of the chapter, but bear with me; I have a point, I promise.

Now, Robert FitzRoy wasn't a particularly bad man by the values of his time, although given that he was born in 1805 that probably isn't saying much. He went to church; loved his country and his family; and he wasn't even that cartoonishly racist as Victorians go – which is to say, he was *astonishingly* racist by modern standards, just not in exactly the way you might expect. He wasn't that bothered by slavery, for example, but not because he thought certain races of people deserved to be exploited for free labour – he only thought certain *cultures* needed to be educated out of people. Better?

OK, it's not all that much better. But FitzRoy was actually seen as dangerously progressive for his time – to this day, he's remembered in New Zealand as a white governor willing to side with the Māori people over European settlers. This happened on not one but *two* occasions, which is probably some kind of record as far as these things go, and actually made him quite a few enemies both in his new home and back in London.

He was also an accomplished scientist, whose legacy can be measured in thousands, if not millions, of lives saved. Appalled by the massive loss of life at sea due to the Victorian inability to predict weather conditions, FitzRoy developed new ways to forecast storms, gifting the world something that his contemporaries thought was impossible but we take for granted to such a degree today that you've probably never even thought of the fact that it had to be invented at some point. He even coined the word 'forecast', in an effort to head off accusations labelling him as some quack, claiming to 'feel it in their bones' or something.

And, in 1865, he took a razor to his own throat, and died by suicide. But while it may have been his own hand that did the deed, there's some who would say he was killed by another man: a scientist he had met more than three decades earlier on a boat named after a dog.

The year was 1831, and Charles Darwin was a fresh-faced twenty-two-year-old with a bachelor's degree and a soft spot for biology. He had been pleasantly surprised by his recent academic performance: he had placed tenth out of 178 students in the final examination for his degree, and he was now one step closer to his planned easy life as a church minister in the English countryside.

First, though, he was going to go on an adventure: a two-year-long voyage around the southern hemisphere on the HMS *Beagle*, captained by FitzRoy. Darwin's mentor at Cambridge had put his name forward for the position of onboard 'gentleman naturalist', which is the kind of rank your crew can include when you're a ridiculous Victorian Tory with a marquis on one side of your family tree and a duke on the other, as FitzRoy was.

FitzRoy had a real reason for taking Darwin on board with him, though: he was there for FitzRoy's safety. Not from the sea – FitzRoy was already a seasoned naval officer who had been part of a crew conducting hydrographic surveys* of Tierra del Fuego, around the southernmost tip of South America. In fact, Darwin was there to protect FitzRoy from himself – his uncle had taken his own life less than a decade beforehand; the previous captain of the *Beagle* had killed himself while in command; and FitzRoy was scared that two years at sea, forbidden by Victorian etiquette from talking to anyone below him in rank, would send him off the same way.

For a while, it worked great. 'He is a very extraordinary person,' Darwin told his sister one year into the voyage. 'I never before came across a man whom I could fancy being a Napoleon or a Nelson.'

But the cracks in the relationship soon started to show. Darwin was a Whig – which is to say, basically left-wing, if the left wing was ruled entirely by upper-class old dudes who all know each other from school (hard to imagine, I know) – and FitzRoy was a rabid Tory; Darwin was a quiet, humble guy who didn't want to cause a fuss, while FitzRoy was so known for his random and extreme outbursts of anger that his crew called him 'hot coffee'. Slavery was a huge point of contention: Darwin opposed it without exception, while FitzRoy, if you recall, didn't think it was all *that* bad – he even told Darwin a story about a guy

---

* And in true Victorian tradition, some light kidnapping and quote-unquote 'civilisation' of native people. The crew of the *Beagle* claimed that some of the native Yaghan people had tried to steal one of their boats, and once the suspects were detained, they decided they couldn't really be bothered to take them back to shore, so instead they gave them English names and clothes, which must have been something of a punishment to them in the age of corsetry and stiffened collars, and tried to 'educate' and Christianise them.

There were four Yaghan abductees, though one died of smallpox soon after his arrival in England. The other three were named Yok'cushly and O'run-del'lico, a boy and a girl renamed Fuegia Basket and Jemmy Button by the crew after the things they were 'bought' with – which is, if nothing else, just incredibly tacky on the crew's part – and a man named El'leparu, who they renamed York Minster, after York Minster.

The three became quite the celebrities in England, even meeting the King and Queen at one point. It seems they were less impressed with the UK, though: they were returned to their homeland the following year in the hopes that they would teach their families and communities the ways of Brits, but within just a few months had fully embraced their former Yaghan traditions.

he knew who asked all the humans he was keeping enslaved whether they wanted to be free and they said no. Darwin immediately said what you're probably thinking right now, which was to ask whether FitzRoy honestly thought a slave was going to tell their enslaver their true opinions on slavery, to which FitzRoy reacted by flying into a rage and banning Darwin from his sight for ever, which was to say, for a few hours before the crew embarrassed him into forgiveness.

Despite having to deal with FitzRoy's ... well, let's just say with FitzRoy, that trip would change the course of Darwin's life and legacy. His time on the *Beagle* – it lasted five years, in the end, more than twice the planned duration – kickstarted his understanding of evolution, in turn transforming the way we understand science and nature on a fundamental level and turning a would-be village parson into a figure who, if you believe some of the more extreme fringes of Christianity, was nothing short of a spokesperson for Hell itself.

But more importantly than all that, it *really* broadened his palate.

See, Darwin was a scientist, but he was also a Victorian, and that meant living in a world where your wallpaper was poison, your baby was on smack, and your MP was voting through measures to genocide the Irish. Compared to all that, why *not* chow down on a few things we might think of as a bit strange today?

Take barn owl, for instance – a bird whose taste Darwin described as 'indescribable'. Joining Hedwig on her journey down Darwin's gullet was a bittern – a type of heron – and a hawk, as well as squirrels and rodents of various stripes, all of whom he ate while still an undergrad at Cambridge. As a child, and even a teenager, he would often try to eat beetles he found in the garden, recording on one occasion his disappointment when a bombardier beetle escaped from his clutches before he could get a taste.*

---

* You can sort of understand his reaction: bombardier beetles are very rare in the UK. But as any toad unlucky enough to have swallowed one of them – only to be quickly seared from the inside by a loud explosion of hot acid from a bug butt – can tell you, putting a bombardier beetle in your mouth is a very bad idea.

So if that's what he was like *before* his job description was essentially 'find and catalogue strange animals', you can imagine the delicacies that were in store for him once he joined the *Beagle*.

For the crew, travelling round South America and the Galápagos, every day must have been a culinary adventure. Darwin served his shipmates pampas deer and llama, Andean condor, and foxes; iguanas, he said, were 'hideous-looking' with a 'singularly stupid appearance', but nevertheless made a delicious meal 'for those whose stomachs soar above … prejudices.' Puma is 'remarkably like veal in taste', he recorded, and Armadillo tastes like duck, bizarrely. Meanwhile, 'the very best meat [he] ever tasted' was an unnamed South American rodent – modern scholars think it was probably an agouti, an animal that looks like a capybara had a baby with a squirrel, smells like ass, and, apparently, tastes divine.

In fact, Darwin's gut sometimes straight-up held back scientific discovery. Some of the breakthroughs we give Darwin credit for today were really less the result of 'diligent field work' so much as 'oh my God, Charles, stop eating that and send it to the Natural History Museum, it's a brand-new species'. The lesser rhea, for example, is also known today as 'Darwin's rhea', because the first known specimen to be sent to London came from Darwin's leftovers – after months of hunting the bird in order to record and classify it for science, he finally caught one only to misidentify it as something else and give it to the ship's cook to serve to the crew for dinner. Darwin didn't even notice the mix-up until everybody was merrily chomping away and asking for seconds, at which point he ran around the kitchen and mess hall like a mad thing, demanding everybody's plates and collecting up what he could of the now ex-bird so that he could send it back to a taxidermist in London. Which means that, yes, the first example of a lesser rhea in England was essentially reconstructed from stew.

Then there were the giant tortoises. These days the gigantic reptiles are practically inseparable from the man himself in the public imagination, given their key role in the development of Darwin's theory of evolution, but at the time Darwin saw giant tortoises less as a charismatic mascot and more like an enormous walking lunchbox. On this, at least, Darwin wasn't alone. The giant tortoises that populated the Galápagos were 'extraordinary large and fat, and so sweet, that no pullet eats more pleasantly', according to the seventeenth-century pirate William Dampier, and around twenty years before Darwin and the *Beagle* made it to the islands, the US Navy captain David Porter wrote that 'after once tasting the Galapagos tortoises, every other animal food fell off greatly in our estimation'.

Thousands and thousands of the creatures were taken from their homelands every year because of this unfortunate trait. It's almost ironic, really, that the giant tortoises were the thing that finally convinced Darwin of the truth of evolution, because they're pretty much the perfect argument in favour of an intelligent designer: they taste delicious, they can live quite happily on ships so you don't need to worry about the meat going rotten, and they even come with their own drink – 'the inhabitants [of the Galápagos islands] … drink the contents of the bladder if full', Darwin recorded, noting that 'in one I saw killed, the fluid was quite limpid, and had only a very slightly bitter taste'.

Best of all, they're good for a laugh. 'The inhabitants believe that these animals are absolutely deaf; certainly they do not overhear a person walking close behind them,' recalled Darwin. 'I was always amused when overtaking one of these great monsters, as it was quietly pacing along, to see how suddenly, the instant I passed, it would draw in its head and legs, and uttering a deep hiss fall to the ground with a heavy sound, as if struck dead,' he wrote. 'I frequently got on their backs, and then giving a few raps on the hinder part of their shells, they would rise up and walk away … but I found it very difficult to keep my balance.'

So delicious and convenient were the beasts that even by Darwin's time – a good 300 years after the giant tortoises had first been discovered – there was still no scientific name for the Galápagos tortoise, as no specimens had managed to make it to the West on board a ship without being eaten en route. The voyage of the *Beagle*, despite ostensibly being a journey of discovery, made absolutely no headway in this direction either: they left the islands with more than thirty tortoises on board, and literally all of them had been served up for dinner by the time the ship got back to England.

Sadly for the tortoises – but luckily for the inevitable march of scientific progress – they didn't need to be alive to spark Darwin's imagination, and, in 1859, Darwin published the book that made him a bona fide superstar.

It was called *On the Origin of Species by Means of Natural Selection, or the Preservation of Favoured Races in the Struggle for Life*, but since that's a ridiculous number of syllables to have as the title of a book, we generally know it today as the *Origin of Species*. FitzRoy, by now a full-on fundamentalist, was appalled – both at Darwin for popularising what was clearly heresy, and at himself for enabling it by inviting Darwin on his boat all those years ago. He started turning up at scientific debates on evolution and denouncing his erstwhile shipmate, holding a Bible aloft and commanding the audience to accept God's word over Darwin's.

But, by now, FitzRoy, once a highly respected scientist and naval officer, was something of a laughing stock among British society. It was the 1860s, of course, so it wasn't for the reason you'd expect – sure, some people thought his ardent anti-evolution campaigning was a bit weird, but the reason he found himself lampooned in *Punch* and criticised by his peers and the public was actually his weather forecasts. Torn between sceptics who still rejected the idea that predicting the weather was possible and believers who thought he was an idiot who got the forecast wrong too often – which seems fairly harsh considering he was the only person in the world who had worked this shit out – and wracked with guilt over what he saw as his part in the downfall of Christian belief, FitzRoy finally succumbed to the fate he had feared so long ago. He sent in his final forecast – thunderstorms across London – and killed himself the next day.

Of course, that wasn't the end of Darwin's story. He would live for nearly twenty more years, getting only more famous and respected. But just as FitzRoy wasn't derided for what you think, so Darwin was celebrated for something potentially unexpected – not evolution, but earthworms.

After the small diversion that was revolutionising the way the entire world thought about nature and our place within it, Darwin was going back to his roots: creepy crawlies. He ran dozens of experiments on worms, some lasting years, to prove important hypotheses like 'worms eat dirt' and 'worms don't care if you shout at them, try it, you'll see'. He even lay out a collection of the little wiggly boys on a table and got his son to blast a bassoon at them to see if they reacted – perhaps he was hoping for them to break out into a polka or something, but it didn't work. He therefore concluded that worms were deaf, which both makes sense (because they have no ears) and is completely incorrect (recent research has found that a worm's entire body acts as an eardrum.)

To modern eyes, *Worms!* doesn't really seem like a worthy sequel to something as utterly ground-breaking as the *Origin of Species*, but in fact Darwin's sophomore offering – which was actually named *The Formation of Vegetable Mould Through the Action of Worms, With Observations on Their Habits*, because why screw with a winning title formula – sold even better than its predecessor.

In the end, though, there was no escaping the curse of the *Beagle*. That fateful voyage had already seen off FitzRoy, and, in 1882, it may have been what finally killed Darwin too – the doctors at the time diagnosed him with blood clots in his heart, but modern scholars have suggested he might have actually been suffering from undiagnosed chronic Chagas disease.

Chagas is a quiet little infection, contracted from the bite of a 'kissing bug' – an insect that Darwin had fallen foul of in 1835, when the *Beagle* was in Argentina. When left untreated, it can lie low for decades, never causing a noticeable problem until, suddenly, it causes your heart to fail. When that fate came to Darwin, he spent his last days surrounded by his wife and family,* telling them that he was 'not the least afraid of death', and 'it's almost worthwhile to be sick to be nursed by you'.

And then, the father of evolution died. And it happened in the most fitting, karmically balanced way possible: thanks to a beetle that bit back.

---

* This was especially easy for Darwin, as he married his first cousin.

19.

James Glaisher, the Victorian Weatherman Who Nearly Became an Astronaut

The Victorian era was a weird time. Like, society was advanced enough to know that germs cause disease, but hadn't yet worked out that simply masking the taste of milk going sour doesn't actually mean the deadly tuberculosis bacteria was gone. We had figured out indoor plumbing, but not well enough that there weren't epidemics of exploding toilets every so often. It was kind of like that awkward stage in early puberty, where you've got boobs but don't yet know what to do with them, you know? Except applied to an entire world population's understanding of science.

And nowhere was this mismatch between engineering know-how and common sense wider than when it came to getting around. So, for instance, Victorians *could* take a train around the country, but they were convinced that doing so would leave them both insane and infertile, which, as anybody who's ever travelled during rush hour on a British train can tell you, is only half true. Short of a train, they could ride a bike, except that they thought cycling was basically one step away from moral anarchy, suitable only for ugly spinsters or, worse, feminists.

Most exciting of all: for the first time in known history, people were taking to the skies. Not in any particularly useful way, mind you – we were still a long, long way away from hopping in a jumbo jet after breakfast in London and arriving in Los Angeles in time for dinner. But if you were rich enough, with strong nerves and spare time – or failing that, one of those olde timey circus performers with a leotard and a great big handlebar moustache – you *could* be one of the very few people to experience the world from 1,000 feet above. All you needed was a balloon.

The first ever hot air balloon ride had been carried out in 1783, in France. Standing on a platform suspended underneath a paper and silk balloon, Jean François Pilâtre de Rozier and François Laurent, Marquis of Arlanders, had ascended 500 feet in the air over Paris, hand-feeding fuel into the fire above their heads to keep themselves airborne. They landed, twenty-five minutes later, in a vineyard more than five miles away, where legend has it they passed out bottles of champagne to the local peasant farmers so as to prove they weren't demons.

In other words, ballooning wasn't some new or untested phenomenon in 1862, when James Glaisher made the ascent that would make him famous. It wasn't even untested by him; he was one of a pioneering new crop of meteorologists who saw ballooning as an exciting new way to understand the Earth's atmosphere, and he had already put the technique to the test quite a few times by this point.

So it must have been pretty embarrassing for him when, on 5 September that year, he hopped into a balloon with three pigeons, a pilot named Coxwell, and a collection of specially designed cutting-edge scientific instruments, and almost immediately shot up into the stratosphere – literally, not figuratively, though the name for that part of the atmosphere hadn't yet been invented at the time – nearly killing everybody on board with a double attack combo of altitude sickness and decompression sickness at the same time.

James Glaisher isn't one of the more recognisable names in this book, which is kind of incredible, because he wrote more than 100 books and papers on multiple areas of science, from astronomy to number theory. He was a leading meteorologist – perhaps *the* leading meteorologist of the time, as he was Superintendent of the Department of Meteorology and Magnetism at the Royal Observatory, Greenwich, for more than three decades. He was a founding member of the British Meteorological Society and its president later on, after it moved up in the world and became the *Royal* Meteorological Society; he was the first guy to say 'hey, maybe we could get lots of people measuring the weather all across the country and compile it all into one weather *map*', and he would personally travel all around the UK as well, just checking up on his outposts to make sure everything was ticking along nicely and accurately.

And that's not all. Outside of meteorology, he was also at various times president of the Royal Microscopical Society, the Royal Photographic Society, and the Aeronautical Society of Great Britain; he was Chairman of the Executive Committee of the Palestine Exploration Fund, and, of course, a Fellow of the Royal Society. And to top it all off, he was rude, bad at following rules, and he literally married a child when he was thirty-four. Basically, the kind of person you sometimes wish really would accidentally fly off into space in a hot air balloon, if only to let you get a word in edgeways.*

Luckily, that was his goal too. 'I was often compelled to remain sometimes for long periods, above or enveloped in cloud,' he wrote in his memoir, *Travels in the Air*, recalling his younger days working in a markedly more grounded position as an assistant cartographer for the Ordnance Survey of Ireland.

'I was thus led to study the colours of the sky, the delicate tints of the clouds, the motion of opaque masses, the forms of the crystals of snow,' he continued. 'On leaving the Survey, and entering the Observatory of Cambridge, and afterwards that of Greenwich, my taste did not change. Often between astronomical observations I

---

* That word presumably being: 'Ew, dude, she's fifteen'.

have watched with great interest the forms of the clouds, and often, when a barrier of cloud has suddenly concealed the stars from view, I have wished to know the cause of their rapid formation, and the processes in action around them.'

Of course, we all drift off into daydreams at work now and then; what very few of us do is convince our employer to fund an expensive and dangerous series of experiments more at home in a sci-fi novel than an observatory to make it happen. But sailing in the 'aerial ocean', as he poetically described the skies, was Glaisher's dream, and so he set out his case to the British Association for the Advancement of Science: he would, with the help of expert balloon pilot Henry Tracey Coxwell, a man most famous at the time for ballooning above Berlin and chucking a bunch of small bombs out of the sky – not for any particularly belligerent reason, that was just the kind of thing people had to do for fun before the invention of the TV – voyage into the unknown world above the clouds, and bring back to Earth the knowledge he discovered there.

He certainly knew how to sell the idea: his major promise was that, by better understanding the atmosphere thousands of feet above the ground, he would bring much-needed clarity to a burgeoning new area of science: weather prediction. Ballooning was, then as now, firmly established as a novelty – suitable for circus performers and champagne 'picnics in the air' maybe, but not for upstanding men of science, who saw their position not just as leaders intellectually but morally too. But if ever there was a way to convince a bunch of stuffy English guys to give you a bunch of money, it's to tell them you're going to open up a new avenue of awkward small talk for parties, and, somehow, Glaisher persuaded the Association that *his* jaunts above the clouds would be different from the ones associated with trapeze artists (or, worse, the French). *His* balloon rides, in fact, would be downright beneficial for the moral character of the nation.

The first time James Glaisher went up in a balloon was meant to have been 30 June 1862. From the very start, the universe was clearly trying to signal that it was a bad idea: it was a horrible day, not just too windy to go up in the air, but also so gusty that *the balloon actually split open* before lift-off. Now, clearly, that would be a problem under any balloon-related circumstances, but it was even worse for Glaisher and Coxwell, since The Mammoth – so called because it was the largest balloon known up to that point – was not a hot air balloon, but a gas balloon.

See, this was a good seventy-five years or so before the world had heard of things like 'global warming' or 'the Hindenburg disaster', so the standard way to get in the sky was to burn a whole bunch of coal until it started releasing hydrogen. When you'd created enough of this, the lightest gas in existence, to counteract the combined weight of a basket, a couple of humans, a bunch of scientific equipment, and an undetermined number of pigeons, you'd shove it all inside a balloon made from silk or, in the case of The Mammoth, 'American cloth', and off you'd fly.

So when the balloon split open that day, it wasn't just a case of sewing it back up and re-lighting the fire. The assembled academics and balloonists had to find a way to replace, with almost zero notice, approximately 60,000 cubic feet of hydrogen – to put it in modern terms, about enough to fill six two-storey new builds.

Lucky for them, then, that they were located right in the middle of the Victorian era, when dubious environmental and human health practices were right at their heyday, and a passing industrialist named Mr Proud was able to provide them with the required gas.

It took seventeen days in all to repair the balloon and secure the replacement hydrogen – actually, it should have taken longer, but 17 July was the last day Glaisher was available for the flight, so seventeen days is what it took. And if the gang were hoping that this second attempt would be more in tune with the will of the cosmos, they were soon to be disappointed: 'the weather increased in badness,' Glaisher later recalled, 'and if it had not been for the already great loss of time ... we should not have set [off].'

But the problems weren't limited to the gales. Not only was the weather full of 'badness', but – and despite the extra gas Mr Proud had provided – the balloon wouldn't fill up. 'It seemed as if the operation would never be completed,' Glaisher wrote. Meanwhile, he was busy trying to calibrate the scientific instruments and doodads he was hoping to bring up with him, which he couldn't because the balloon was too unstable, and 'it was impossible to fix a single instrument in its position before quitting the earth,' he said.

'The state of affairs was by no means cheering to a novice who had never before put his foot in the car of a balloon,' he wrote, in typically understated Victorian fashion.

It took nearly five hours before the balloon was declared ready to fly – by which I mean, it took nearly five hours before the group said, 'ah, fuck it, this'll do, I'm bored', and decided to see if they could give up yet. They let the balloon go, at which point it immediately started ploughing along the floor, dragging the upended basket behind it like some kind of unstoppable and extraordinarily jolly M1 Abrams tank. '[It] would have been fatal had there been any chimney or lofty buildings in the way,' Glaisher recorded.

The team continued this slow-mo Benny Hill sketch for a couple of minutes, just trying to regain control of the renegade balloon, until eventually, with Coxwell at the … helm? I guess? Glaisher finally, *finally*, realised his dream of flight. Together, the pair ascended more than five miles into the sky, and if that sounds pretty far to go on what is obviously the most cursed day for balloon flying in history to you, it's because it is. It is, in fact, approximately four and a half miles higher than a standard, calm skies, no-fuss balloon flight will take you today.

'Mr Coxwell – himself, of course, a thoroughly experienced aeronaut – was loud in his praises as to the perfect coolness and self-possession shown by Mr Glaisher, who had not previously been up in a balloon,' recorded *The Times*, many years later and making no mention of the absurd comedy of errors that made that sangfroid impressive in the first place. It was kind of a weird thing for Coxwell to have been wowed by, really, since the stress of the situation would have been arguably even worse for him – after all, he'd never been that high either, and even worse, he knew why: all previous attempts to do so had resulted in not just failure, but also a completely fucked-up balloon. Plus his professional reputation (as well as his life) was on the line: this was *his* balloon; *his* flight; *his* middle finger being thrust right up in the faces of various weather- and hubris-related gods, and *he* knew the stakes in a way Glaisher would have been happily ignorant of.

Still, the pair rose quickly, reaching a height of nearly 8,000 feet within the first ten minutes or so. 'A most magnificent view … presented itself,' Glaisher noted, although he didn't see it at all because he was still pottering about trying to get his instruments all calibrated. By the time he managed to set them up, the balloon was almost two miles high,* and the Earth was completely hidden by a layer of cloud.

The temperatures got colder as they continued their ascent; by about three miles up it had sunk to below freezing.

Another half a mile up, and Glaisher was starting to feel the effects of altitude sickness. 'My pulse beat at the rate of 100 pulsations per minute,' he recorded; 'it was with increasing difficulty that I could read the instruments; the palpitation of [my] heart was very perceptible [and my] hands and lips assumed a dark bluish colour,' he wrote.

Another half a mile: 'I experienced a feeling analogous to sea-sickness, though there was neither pitching nor rolling in the balloon; and through this illness I was unable to watch the instruments long enough to lower the temperature to get a deposit of dew.'

Still, they continued. Glaisher watched helplessly as his instruments spat out first readings that seemed to be nonsense, and then stopped working altogether. He struggled to breathe; he heard

---

* Glaisher does mention that they 'had both thrown off all extra clothing', but this fact is unfortunately almost certainly unrelated to their twice-over membership of the Mile High Club.

the ticking of his watch grow deafeningly loud, but he could no longer read it.

And then, eventually, Coxwell decided it was time to descend. They had travelled too far East, he reckoned, and despite starting their journey in Wolverhampton, they were now in danger of flying out across the North Sea. And so down they went – and by that, I of course mean that they flirted with death again by falling out of the sky at a speed of more than 45 miles per hour.

'From the rapidity of the descent the balloon assumed the shape of a parachute,' Glaisher wrote. 'We [had] collected so much weight by the condensation of the immense amount of vapour through which we passed, that … we came to the earth with a very considerable shock, which broke nearly all the instruments.'

The entire flight took little more than two hours, and it's hard to imagine what else could have gone wrong in that time: the scientific apparatus had only served to get smashed up at the end; they'd gone up so high and so fast that Glaisher had apparently felt his soul leaving his body, and then come down so fast they'd injured almost everything on board; and to top it all off, they were now about 50 miles away from where they had started, alone, in Rutland.

And yet all of this had been laughably smooth compared to the journey the two would take a couple of months later.

Oh, you'd forgotten, had you, in amongst all of that chaos, that this wasn't actually the famously disastrous flight? No – unbelievable though it may sound, not only were Glaisher and Coxwell not put off ballooning for life after this initial escapade, but they apparently managed to get even worse at it.

It was another horrible day when the pair set out, accompanied by their new equipment and six pigeons – cold and misty, and no good for ballooning, not least because they could hear people firing guns somewhere below them, which is never a good sound when your life is dependent on the integrity of a balloon. Almost immediately, things started getting – how can I put this – *foreboding*: 'beneath us lay a magnificent sea of clouds, its surface varied with endless hills, hillocks, and mountain chains, and with many snow-white tufts rising from it,' Glaisher recalled. 'I here attempted to take a view with the camera, but we were rising with too great rapidity and revolving too quickly to enable me to succeed.'

Within twenty-five minutes of their launch, they were facing below-freezing temperatures and could no longer see the ground. They were three miles up, and the equipment had already started to break: it had got too cold too quickly for the wet-bulb thermometer to be of any use. He chucked his first bird out of the basket, casual animal cruelty being a mainstay of science at this point, and watched as she opened her wings and 'dropped like a piece of paper' to the ground.

The truth was, that pigeon was the luckiest passenger on board. As the balloon climbed further, Coxwell started struggling to breathe; they carried on their ascent.

At four miles, even more instruments were breaking under the pressure, or, to be more accurate, the *lack* of pressure, of being so high up. A second pigeon was thrown from the balloon; she struggled in the air, flying in tight circles and dipping slightly each time, but she survived.

As they approached the fifth milestone, Glaisher launched another bird from the balloon: she fell 'as a stone', he noted. That was the extent of his scientific recordings at this point, because he was starting to feel the 'balloon sickness' himself. He could no longer read his scientific instruments; he couldn't make out the details on his watch or his thermometers; after a while, he couldn't even move.

'I laid my arm upon the table, possessed of its full vigour, but on being desirous of using it I found it powerless – it must have lost its power momentarily; trying to move the other arm, I found it powerless also,' he recorded. 'I tried to shake myself, and succeeded, but I seemed to have no limbs.'

He tried to check the barometer, to get an idea of how high they had risen, but he couldn't lift his head to read it. He tried again, and collapsed backwards, his head landing on the edge of the basket. He felt sure he was going to die in that balloon: 'I dimly saw Mr Coxwell, and endeavoured to speak, but could not,' he recalled. 'In an instant intense darkness overcame me ... I thought I had been seized with asphyxia, and believed I should experience nothing more, as death would come unless we speedily descended.'

That would be the last thought he had before he blacked out completely. The last observation he managed to record put the pair at more than 29,000 feet in the air – at least five and a half miles high – but that had been a few minutes ago already. They needed to go back down.

There was just one problem. They couldn't.

Modern hot air balloon experiences don't fly when the weather's bad. They don't fly when it's too cold, or too rainy; they don't fly when there are strong or unpredictable winds. We should be thankful for that, because if they *did* fly under those conditions, we might end up experiencing the same problem that Glaisher and Coxwell faced all those years ago – albeit at a small fraction of their altitude. See, as the wind had jostled and spun them when they first left the ground – when Glaisher had been so peeved that he couldn't take a photo of the clouds as they passed – the valve-line had got caught up in the mass of ropes above their heads.

That valve-line was the only way the two men had to control the gas flow into the balloon. The only way they could possibly make a descent. And it was tangled, unreachable, above their heads, while they were trapped in the basket and rapidly losing control of their senses.

As Glaisher lost consciousness, the last thing he saw would have been the dim shape of Henry Coxwell, struggling to breathe and barely in control of his own limbs, clambering out of the basket and up, into the rigging, to release the valve-line and save the pair. But his hands were frozen from the incredible cold, and he wouldn't be able to use his arms for much longer, and so, in a move that any Hollywood action star would be proud of, he grabbed the valve-line with his teeth, and yanked his head back as hard as he could.

It worked. With a few more pulls, the balloon started moving downwards. Coxwell had saved them both – as well as one lone pigeon, the only one to survive the ordeal, who ended up so traumatised by the whole thing that she refused to fly after the balloon landed, instead clinging to Glaisher's hand for a full fifteen minutes.

And who can blame her? All her friends were dead or disappeared, and she'd been taken to heights that wouldn't be reached again until the dawn of commercial passenger jets. It's estimated today that the balloon went about thirty-six or thirty-seven thousand feet in the air – around seven miles high, way above even the tip of Mount Everest. With nothing but their Victorian workwear and, in one case, a full set of feathers, to shield them, it's not surprising that Glaisher blacked out for a good seven minutes. It's honestly amazing they survived at all.

But survive they did, and we should give them this at least: their ridiculous adventure was not in vain. By shooting themselves so far into the sky multiple times, they had discovered, albeit by accident and occasionally while unconscious, the tropopause, the first indication of the existence of the stratosphere, and the foundation of what we now know as atmospheric science. They worked out how raindrops formed, and found the first evidence for vertical wind shears – the phenomenon where wind speeds and directions change with altitude rather than latitude – though they couldn't measure it at the time, on account of being passed out. And those seemingly nonsensical temperature readings that Glaisher's instruments started spitting out before they gave up on that first flight – well, now we know what caused it. It's called temperature inversion, and it's the intuitively impossible effect where temperatures get hotter as you go further away from the Earth. It turns out, things get a little weird when you're at an altitude no mortal should ever experience.

Glaisher wanted to change the discipline of meteorology for ever, and he absolutely did – even if he nearly ended up as the Earth's first man-made satellite in the process. Balloon observations like his are still carried out today, except that we take the sensible precaution of not actually putting anything alive in there these days, and there are even some people who think that ballooning to such stupid heights may be the future of space travel – though since we'll all apparently pass out at seven miles up, I can't see how it'll be all that impressive.

The story has a delightfully Victorian ending, by the way. Ever the stoic scientist, Glaisher recalled that it was Coxwell's voice talking to him about the need to continue his meteorological observations that woke him up. 'I have been insensible,' he admitted to the man who just saved his life, before finding his pencil and resuming their scientific work.

'You have, and I too, very nearly,' Coxwell replied. He lifted his now useless hands to show his passenger – they had turned black from the cold and altitude. Glaisher opened a bottle of brandy and poured the liquor over them.

'No inconvenience followed my insensibility,' Glaisher wrote, which must have been nice for Coxwell and his deadened, boozy hands. In fact, the biggest gripe Glaisher had was the fact that they landed in Cold Weston, a near-abandoned village close to the Welsh border, and the pair were forced to walk seven miles back to where they started – a complaint that would ring less hollow if they hadn't literally just done exactly that, but upwards.

Oh – and since you were wondering, no, nearly dying on the edge of space did *not*, in fact, dampen either of the men's taste for ballooning. Over the next four years or so, Glaisher would continue to fly outrageously high in the vaudevillian aircraft – though never again to 37,000 feet – often with Coxwell as his pilot. Which really just goes to show: if you want to revolutionise a whole field of science, it helps if you're really fucking bad at learning from your mistakes.

20.

Sigmund Freud Used Cocaine So Much He Thought Numbers Wanted to Kill Him

Psychiatry, you'll be pleased to know, has come a long, long way since Johann Reil first looked at a cat and decided it would probably make a good anti-anxiety medication.* These days, the standard treatment for most mental health problems is more likely to involve talk therapy than musical pet torture – and, for that, there's pretty much just one person to thank: Sigmund Freud.

So influential was the Viennese 'father of psychoanalysis' that it's amazing more movies haven't been made about him. But then again, why would you? After all, the perfect Freud biopic already exists: it came out in 1983, and it's called *Scarface*.

Now, *Scarface*, if you don't know, is a movie about Tony Montana, a man who starts the film as a refugee and spends the next two and a half hours talking about his balls, his honour and all the various vaginas (both real and metaphorical) he'd like to get inside. Tony has clear mommy issues, a relationship with his sister that has

---

* See page 117.

frankly incestuous vibes to it, and a father figure whom he kills then whose widow he marries, which is just about as Oedipal as you can get. Even if you've never seen the film, I bet you can quote it, and most of its most famous lines are already pretty Freudian: you have to assume the man who thought pretty much any mental illness is a result of sexual dysfunction would have a lot to say about Tony's 'all I have in this world is my balls and my word, and I don't break 'em for no one', and he probably would have had a field day with 'say hello to my little friend!'

But none of that is what makes *Scarface* the ultimate homage to Sigmund Freud. It turns out what Freud and Montana would have truly bonded over was their penchant for doing, and pushing, just massive amounts of cocaine.

Of course, Freud started his coke habit about a century earlier than Tony – he first got his hands on what was then a fairly obscure narcotic in the early 1880s. At that point he was working full-time at the Vienna General Hospital, because people in the past were just *way* more chill about certain things than we are, and when he wasn't risking patients' lives by practising medicine while on blow, he was busy love bombing his future wife Martha Bernays with increasingly coke-addled love letters.

'Woe to you, my princess, when I come,' he wrote to her in June 1884. 'I will kiss you quite red and feed you till you are plump. And if you are forward you shall see who is the stronger, a gentle little girl who doesn't eat enough or a big wild man who has cocaine in his body.'

So enamoured was Freud with his two mistresses – Martha and the nose candy – that he did what so many of us fantasise about and merged them together into one super-girlfriend, prescribing his fiancée the drug to 'give [her] cheeks a red colour'. Plenty of biographers throughout the last century have pointed out Martha's significance in Freud's achievements – she's been immortalised in the history books as the ultimate housewife and mother, roles which she was arguably forced into by her husband, who both Martha and Freud himself admitted was a 'tyrant' – but what tends to be left out of those conversations is the idea that it's probably easier to clean house, cook, raise six kids, not to mention *make* six kids, if you're constantly high on mountains of blow.

In Siggy's defence, it's kind of impossible to overstate the extent to which absolutely everybody was high as balls in the Olden Days. Obviously, people have been getting wasted since time immemorial – even chimpanzees have worked out ways to get drunk off tree sap, so it's clearly hardwired into our DNA to party hearty – but it wasn't until the late modern age that humanity's drug use moved from 'social' to 'problematic'. First there was laudanum: a wonder drug that could make you forget just about any pain or illness currently tormenting you, on account of it literally being a mixture of opium and hard liquor. Consequently, it was prescribed for … well, just about any pain or illness currently tormenting a person, including period cramps, colds, heart disease, dysentery, babies' teething pain and, in at least one case,* writer's block.

By the nineteenth century, people were starting to realise opium and laudanum addiction existed and were Bad, probably because just about everyone up to and including the President of the United States knew someone who was completely dependent on the drug – in Honest Abe's case, it was his wife, Mary Todd Lincoln. At the same

---

* This would be Samuel Taylor Coleridge, Romantic poet and friend of Lord Byron, who famously got completely off his face on opium and laudanum one day in 1797, fell asleep and dreamed a 300-line poem about the thirteenth-century Mongol Emperor Kublai Khan. When he woke up he immediately set about frantically writing down this epic vision, appropriately called *Kubla Khan*, but unfortunately about thirty lines in some guy from Porlock – literally the very first 'person from Porlock' – came to visit for an hour or two, during which time Coleridge sobered up too much to finish the tale. For this reason, if you ever read *Kubla Khan*, you will notice three things: firstly, it's only about fifty lines long instead of a few hundred; secondly, the last third or so is in a jarringly different style, almost as if it had been written by somebody else completely; and thirdly, it's prefaced by an introduction from the author complaining about this unannounced, and now infamous, well-wisher.

time, though, the world of narcotics was seeing a massive hard-core-ification: this was the period when shooting up was discovered, and when morphine and cocaine and heroin and even meth first found their way into the local pharmacies.

So, far from being a dirty little secret or blot on his authority, Freud's coke habit put him in the same league as people like Charles Dickens (opium), the Pope (wine laced with cocaine, kept in a flask on his person at all times), the entire German military (a shit-ton of methamphetamines), and literally anybody who ever drank a bottle of Coca-Cola or 7Up (cocaine and lithium respectively).

Freud, for his part, credited cocaine with treating his chronic anxiety, as well as working 'against depression and against indigestion, and with the most brilliant success'. He waxed lyrical about the drug's ability to unlock things in his brain and push them out of his mouth seemingly without end – you know, like coke does – and prescribed it for his patients and friends. In fact, Freud was such a fan of the booger sugar that modern biographers have credited him as being more responsible for cocaine's use as a recreational drug than any other person in history.

And, frankly, the same is true in the other direction, because as much as we think of Freud today as being the quintessential psychiatry guru, really the only legacy he has left today is the blow. Pretty much every other theory and practice he came up with apart from talk therapy as a general concept has now been abandoned. That's right: the Oedipus Complex? Nonsense – Freud made it up because he couldn't cope with the idea that as many of his patients had been sexually abused as told him they had. The Id, Ego and Superego? Basically reduces humanity to mindless animals who can only be civilised through violence. And dream analysis? Sorry, new-age mamas: not only have his theories been roundly debunked, but the examples given in his seminal work on the topic are what modern addiction specialists refer to as 'a using dream' – in other words: it was brought on by a whole load of coke working its way around Freud's brain.

Even his work on cocaine – which is to say, about cocaine, but let's face it probably also *on* cocaine – hasn't held up. He managed to completely miss the fact that the drug has analgesic properties, and was convinced that it was completely non-addictive, a misconception he may genuinely have come by due to never giving his body enough time off coke to start craving it. 'It seems to me noteworthy – and I discovered this in myself and in other observers who were capable of judging such things,' he wrote, 'that a first dose or even repeated doses of coca produce no compulsive desire to use the stimulant further.'

And here's where we come to another similarity between Sigmund Freud and Tony Montana: both of them killed their best friends while high. The only difference is that Tony was arguably more merciful, killing his loyal childhood-pal-cum-sidekick-cum-henchman Manny by shooting him to death for schtupping Tony's sister. Freud's best friend – a fellow Viennese physician named Ernst von Fleischl-Marxow – suffered a much longer, more painful end.

The beginning of Fleischl's downfall began in 1871, when he was just twenty-five. It was his first year as a newly qualified doctor, and, as medical folk are sometimes known to do, he was cutting up a dead body to figure out why a patient died. Somehow, despite a dead guy being pretty much the safest kind of guy to cut into, he managed to end up slicing open his own thumb, and since it was 1871 and germ theory was naught but a glint in Robert Koch's eye, that small thumb cut turned into a big, infected problem. After not too long, the thumb had to be amputated, and then, as if Fate herself had a grudge against Fleischl, even this went wrong, and he was left in agonising and untreatable pain for the rest of his life.

Naturally, he turned to the standard painkiller for the time: morphine – and in a twist nobody could have foreseen, he became hopelessly addicted to the (famously incredibly addictive) drug.

Then, in 1882, he met Freud. And Freud had just the thing to cure his problem.

'A sudden abstinence from morphine requires a subcutaneous injection of 0.1 gram. of cocaine,' Freud advised his friend. 'In 10 days a radical cure can be effected by an injection of 0.1 gram. of cocaine 3 times a day.'

Got that? To cure his morphine addiction, Fleischl was to inject 0.1 grams of cocaine three times a day, for a week and a half. That's the kind of thing scientists do to lab rats when they want to see how quickly they can induce a heart attack. And the worst part was, as you've almost certainly already guessed, it only had the effect of turning Fleischl into a morphine *and* cocaine addict.

Within just a few months, Fleischl's coke use was so heavy that *the drug manufacturer themselves* asked him to report back the experiences from such high doses because they didn't know what it would do to a body, and I'm just going to give you a minute to think over the full implications of that sentence. Freud saw his friend bounce between 'the clearest despair up to the most exuberant joy over bad jokes' and paw at imaginary unseen creepy crawlies running up and down his skin – Freud put this down to withdrawal, but modern users know it as the 'coke bugs'.

Six months later, Freud wrote home to Martha that 'Fleischl looks miserable, more like a corpse' and 'hallucinates constantly and it will probably not be possible to let him remain in society for much longer'. And yet, amazingly, he still thought he had done the right thing for his friend – when a morphine addiction specialist of the time, Friedrich Albrecht Erlenmeyer, reported that prescribing cocaine to his patients had only served to give them another addiction, and that Freud had only given the world a 'scourge of humanity [in] cocaine', Freud essentially replied, 'Nuh-uh! Look at Fleischl! He's not using morphine any more!'

Now, Fleischl was *absolutely* still using morphine. Speedballs – that is, a mix of opiates and cocaine – are famously one of the most intoxicating, euphoric and most of all addiction-forming habits, and I say 'famously' because they've killed off quite a few Hollywood stars over the years. That's because, as a mixture of a potent upper and a powerful downer, it's also one of the most dangerous cocktails you can take: users are more likely to overdose, less likely to be able to quit, and risking death by heart attack every time they line up a hit.

And so, in 1891, at the age of forty-five, Ernst von Fleischl-Marxow died. Josef Bauer, a mutual friend of Freud and Fleischl and renowned physician in his own right, wrote at the end that he 'bemoan[ed] Ernst, as I have done for years, but I cannot say that I bemoan his death ... We all owe a death to Nature, but not suffering, not this pathetic crumbling of such a brilliant personality.'

It was, without a doubt, a tragedy, and the only thing I can say to soften the sting is, well, at least Freud was also suffering because of his miracle no-downside non-addictive drug habit. By the 1890s, he was suffering from chest pain, paranoia, depression and nasal congestion so bad that he had to go under the surgeon's knife to relieve it. He started to become convinced that the universe was sending him messages through the appearance of various numbers, especially sixty-two, which he believed was the age at which he would die.

In fact, he was so high-strung about the idea he was being chased by numbers that, on one holiday to Greece in 1904, he spent the whole time recording instances of the numbers sixty-one and sixty-two cropping up on 'anything that had a number, especially on vehicles', he later wrote to a friend.

When he got to the hotel he was meant to be staying at, he was initially relieved – he had been assigned a room on the second floor. At least, he thought, that meant he couldn't have a room in the sixties. But then the concierge gave him the keys for room thirty-one, which he said was even worse, as it was 'after all half of ... sixty-two.' The holiday was ruined: 'This wilier and nimbler figure proved to be even better at dogging me than the first,' he wrote.

So, after nose surgery, weird-ass dreams, chest pain, depression, a dead best friend, and a cosmic chase across the continent involving a group of numbers which he would later realise were just the digits from his phone number, you'd think Freud would have reached his Rock Bottom by now. But, unfortunately, when you're a nineteenth-century man possessing the title of 'medical doctor' and an unlimited supply of cocaine, there are always further depths you can plumb, and that's why, four years after losing Fleischl, Freud nearly added another body to his running kill count.

In *The Interpretation of Dreams* – widely considered today to be one of Freud's most significant works, even if it did take eight years to sell out its first run of 600 copies – there's a famous example dream known as 'Irma's Injection'. Freud describes being at a party when he's approached by the titular Irma, who complains of feeling like she's 'choking' from 'pains … in [her] throat and stomach and abdomen'. Freud inspects her, and invites his doctor friend to inspect her, and both come to the same conclusion: yep, she's sick. All over. Scabs inside her mouth; bowel obstructions; shoulder problems; you name it, she was suffering it. Freud notes, too, that the assembled doctors are 'directly aware, too, of the origin of the infection. Not long before, when she was feeling unwell, my friend Otto had given her an injection … And probably the syringe had not been clean.'

The hidden meaning behind Irma's Injection was clear, Freud thought: his subconscious was chastising him for a medical mistake he had made back at the start of his career. The fact that the second doctor in the dream told him Irma's infection would clear up on its own was the resolution of that discomfort: 'The dream acquits me of responsibility for Irma's condition, as it refers this condition to other causes,' he wrote. 'The dream represents a certain state of affairs, such as I might wish to exist; the content of the dream is thus the fulfilment of a wish; its motive is a wish.'

But what did Freud wish to be acquitted of? Many commentators, including Freud's friend Max Schur, have pointed to a woman named Emma Eckstein as the true identity of 'Irma'.

There are few real-life stories that scream 'everybody involved in this was high as balls on cocaine' more than that of Eckstein and

Freud. She first sought out his treatment in 1892; she was twenty-seven years old, and suffering from what these days we'd probably think of as pretty normal PMS. In a wildly Victorian move, Freud diagnosed her with hysteria caused by too much masturbation and prescribed nose surgery to treat it.

That probably sounds like a particularly bizarre solution to her 'problem', and that's because it is: Freud was working on a theory at the time, originally devised by his friend and fellow coke-fan Wilhelm Fliess, that the nose could be mapped one-to-one to the genitalia – sort of like reflexology, but even stupider.

Sexual dysfunction, Fliess believed and Freud endorsed, could therefore be cured by burning specific bits of the inside of the nose, a treatment that sounds even worse than having 'too much wanking' in your medical history. And for Eckstein, the proposed treatment was even more squirm-inducing: the surgery was intended to remove one of her nasal bones from deep inside her face.

Are you feeling grossed out yet? Because the next bit is worse: Fliess literally left half a metre of gauze inside Eckstein's face, causing her to bleed so profusely from her nose that she almost died. The operation to remove the gauze – which was carried out by a different surgeon – was so horrific to watch that Freud fled the operating room, and Eckstein ended up permanently disfigured because of the botched operation.

But despite seeing first-hand the cause of Eckstein's brush with death being pulled out of her now permanently caved in face-hole, Freud contrived to place the blame for the whole ordeal at her feet rather than Fliess's. All that blood, he decided, had just been Eckstein wanting attention: 'so far I know only that she bled out of longing,' he told Fliess. 'She has always been a bleeder ... she became restless during the night because of an unconscious wish to entice me to go there, and since I did not come during the night, she renewed the bleedings as an unfailing means of re-arousing my affection.'

It's common to see this episode framed as Freud's 'come to Jesus' moment – the point where he finally realised he needed to kick the coke habit. And yet, about four months after permanently ruining his friend and patient's face, he wrote to Fliess that 'I need a lot of cocaine.'

As the nineteenth century became the twentieth, Freud's celebrity seemed only to grow: he founded the 'Wednesday Psychological Society' for physicians interested in his work, kickstarting the beginnings of psychoanalysis as a discipline – according to attendee Max Graf, 'There was the atmosphere of the foundation of a religion in that room [and] Freud himself was its new prophet.' In 1910, Freud founded the International Psychoanalytical Association, a body that's still active today, and over the next two decades psychological institutions and schools started popping up in cities across the world, from New York to Moscow to Jerusalem.

But all good things must come to an end, and that's especially true when you're an inveterate chain smoker and cocaine user. In 1923, Freud found a growth in his mouth that turned out to be the first sign of jaw cancer. At first, his friends kept the seriousness of the diagnosis from him, worried that the news would make him commit suicide, but before long the growth had grown so large that multiple surgeries were needed. Despite his doctors' advice to quit smoking, Freud never gave the habit up, even when it was so painful to move his jaw that he had to prop his mouth open with a clothes peg.

Then, the backlash started against his work. Actually, 'backlash' is underselling it: the Nazis came to power in Germany in 1933, and Freud's books – works on wishy-washy Bolshevik nonsense like dream analysis, and written by a proudly Jewish man, no less – were among the first to be burned in the streets by the newly empowered fascists.

'What progress we are making,' Freud remarked at the time to friend and biographer Ernest Jones. 'In the Middle Ages they would have burned me. Now, they are content with burning my books.'

Five years later, the Nazis annexed Austria. Freud watched as his friends and relatives fled Vienna for safer cities like Paris and London, but it wasn't until his own daughter was interrogated by the gestapo that he could bring himself to follow them. In June 1938, after months of planning the escape, Sigmund, Martha and their daughter Anna boarded the Orient Express and headed west, arriving as refugees at Victoria station in London two days later.

All four of the sisters Freud left behind in Vienna would later be murdered in Nazi death camps.

Freud enjoyed his fame in London, welcoming such luminaries as Salvador Dalí, Virginia Woolf and H. G. Wells into his Hampstead home. But it wasn't long before the pain in his jaw made life unbearable, and he called on his old friend Max Schur to deliver on a deal made years before.

'Schur, you remember our "contract" not to leave me in the lurch when the time had come,' he said. 'Now it is nothing but torture and makes no sense.'

Schur agreed to help Freud end his life, and so, on 22 September 1939, Sigmund Freud died. He was eighty-three, meaning he'd outrun his cocaine-induced death prediction by more than two decades, which is a win. He had changed the world: he'd given us the idea of boys being in love with their mothers and girls being a mystery wrapped in an enigma wrapped in a diagnosis of hysteria, both of which are ridiculous, and he'd also given us the idea that talking about your problems can be helpful, which isn't. But most of all, he gave us cocaine. Masses and masses of cocaine. A drug which puts colour in your cheeks, treats depression and isn't at all addictive.

21.

Arthur Conan Doyle Gets Pranked So Hard He Claims Fairies Exist

There have been few geniuses throughout history quite as iconic as Sherlock Holmes. After all, Einstein may have come up with two new theories of relativity, but he never came back from the dead after plunging hundreds of metres in a final fight with his evil arch-nemesis. In fact, it's unclear whether Einstein even had an evil arch-nemesis, putting him at a disadvantage to Holmes right off the bat.

The detective's legendary brilliance has inspired countless adaptations over the years, spanning film, TV, comic books, radio and even board games. He's turned up in songs, in video games; he's even blessed the English language with new turns of phrase such as 'elementary, my dear Watson' and 'no shit, Sherlock'. His only real shortcoming is that he is, of course, fictional, and therefore disqualified from true geniushood by virtue of not actually existing.

You'd think, though, that if there were one person capable of matching such a character, it would be his creator. Shouldn't the person who came up with that mind be even more impressive? Why don't we use 'Arthur Conan Doyle' as a shorthand for mental prowess?

Well, it's not that complex, honestly: he just wasn't that smart. Not how Holmes is, at any rate – not cool and calculating and oh-so-sceptical. Because Sir Arthur Ignatius Conan Doyle is most famous today for two things: creating one of the greatest logical and scientific thinkers in the history of literature, and publicly claiming that fairies existed after falling for an embarrassingly obvious prank pulled by two random schoolgirls.

Now, despite the image you probably have in your mind of a man with the middle name Ignatius, Doyle came from surprisingly humble beginnings – his childhood was honestly more *Shameless* than *Sherlock*. Born into poverty in 1859, and with a cripplingly alcoholic father, Doyle and his eight siblings spent many of their formative years separated out across Edinburgh, relying on friends and family to help them get by.

Nevertheless, Arthur spent seven years at an expensive boarding school, which just goes to show that you don't need to be born to rich parents to achieve your dreams as long as you have a handful of wealthy uncles willing to pay for you to attend fancy private schools away from your destitute father instead. It's fair to say he fucking hated his time there – Victorian English boarding schools have never been known for too much love and kindness – and outside of his twin pursuits of writing to his mum and playing cricket, he spent his time there resenting his bigoted peers and getting regularly beaten up by his teachers as part of a ritual known at the time as 'education'.

'Perhaps it was good for me that the times were hard, for I was wild, full blooded and a trifle reckless. But the situation called for energy and application so that one was bound to try to meet it,' he would later write about the experience – and, in any case, he added, 'my mother had been so splendid that I could not fail her.'

Now sixteen, and A Man, he went home to Edinburgh, where his first task was to sign the papers to have his dad sectioned, just in case you thought that traumatic childhood had ended at graduation.

It was finally time for Arthur Doyle to make his own way in the world – so he decided to apply for medical school.

It was a decision that must have surprised his family. There are plenty of what's called 'hereditary occupations' out there, even today – fishing is a big one, as are things like joining the military or working in some legal field. Basically, if your parents have one of those jobs, you're statistically way more likely to go that way yourself.

For the Doyles, it was Art. Arthur's father had been an artist – and actually, he still was, producing some increasingly bizarre stuff from his cell in Montrose Royal Lunatic Asylum – and so had been his father's father, John Doyle. Neither had been particularly good or successful, if we're honest, but if there's one thing the Victorians knew, it was that tradition trumps common sense every day, and so, by rights, Arthur ought to have gone into drawing too.

Instead, he wanted to be a doctor. Inspired not by family, but by a student his mother was renting a room to – once again, I feel like I should point out just *how* divorced his dad was from the rest of the family – he applied to the University of Edinburgh to study medicine.

It was an auspicious decision. If Arthur had decided anything else – a different university, or a different course, or even just delayed his application by a couple of years – then we likely wouldn't have one of the most recognisable and successful fictional characters known to literature today. I mean, hell – he might have *actually* been a doctor for a living, and what a loss that would have been to the world.

See, it was at the University of Edinburgh that Conan Doyle, now newly if subtly renamed, met his muse: a man named Joseph Bell. He was one of Conan Doyle's lecturers, and a surgeon by profession, but that wasn't what made him famous. His party trick, if you can use that term for a mid-lecture demonstration, was to visually dissect a patient's life and circumstances based on their clothing, demeanour, accent and so on, and I'm sure that sounds familiar to anybody who's read or seen or heard of Sherlock Holmes, because it's exactly what he does too.

'Dr Bell would sit in a receiving room, with a face like a [Native American, but in a term that a Victorian English dude would use, if you know what I mean], and diagnose people as they came in, before they even opened their mouths,' Conan Doyle once told an interviewer of his inspirational teacher. 'He would tell them their symptoms, and even gave them details of their past life, and hardly ever would make a mistake.'

Compare that with his hero Holmes's appraisal of a woman who 'had plush upon her sleeves, which is a most useful material for showing traces. The double line a little above the wrist, where the typewritist presses against the table, was beautifully defined ... observing the dint of a pince-nez at either side of her nose, I ventured a remark upon short sight and typewriting, which seemed to surprise her.'

And: 'she had written a note before leaving home but after being fully dressed ... both glove and finger were stained with violet ink. She had written in a hurry and dipped her pen too deep. It must have been this morning, or the mark would not remain clear upon the finger. All this is amusing, though rather elementary, but I must go back to business, Watson.'

In fact, Holmes ended up being such a blatant rip-off of Conan Doyle's old lecturer that when his old university friend, Robert Louis Stevenson – who was also a novelist rather than a scientific man, which might make a person wonder exactly what was going wrong in the university's STEM departments at the time – first read the character he felt compelled to write to Conan Doyle to ask 'can this be my old friend Joe Bell?'

But, of course, we're getting ahead of ourselves. The Arthur Conan Doyle who was studying medicine under Bell had dipped his toe into the waters of fiction, but he was still only writing short stories for magazines – and he still intended on becoming a doctor in the end. And so, after graduating from Edinburgh, Conan Doyle didn't head to a publisher or a literary agent; instead, he moved to the south coast of England with the equivalent of about £1,000 total to his name, and set up a medical practice.

It was here that he began his lifelong habit of failing, hard, at literally almost everything he tried. For weeks he had no patients at all – and when someone eventually did knock at his door, it wasn't a customer but a debt collector.

'Through the glass panel I observed that it was a respectable-looking bearded individual with a top-hat,' he later wrote in *The Stark Monro Letters*, a work of 'fiction' that was, in reality, more like a diary that just had all the names and places changed.

'It was a patient. It MUST be a patient!' he wrote. '[I] waved him into the consulting-room ... He seated himself at my invitation and gave a husky cough.'

Showing the kind of deductive genius that would eventually make him famous, Conan Doyle immediately diagnosed the visitor with a bronchial problem, and that probably would have been quite impressive had the man not in fact simply been sent by the gas company to collect on a bill – one that wasn't even Conan Doyle's to begin with.

'You'll laugh ... but it was no laughing matter to me. He wanted eight and sixpence on account of something that the last tenant either had or had not done. Otherwise the company would remove the gas-meter,' Doyle wrote. 'How little he could have guessed that the alternative he was presenting to me was either to pay away more than half my capital, or to give up cooking my food!'

And genuine customers, when they did finally turn up, sometimes couldn't pay him enough to cover his own costs – as he wryly complained in a letter to a family friend, 'if I got more patients I would have to sell the furniture'. By the end of his first year as a physician, he had earned so little that his tax return was literally rejected for being suspiciously low – it was returned to him with a note reading 'most unsatisfactory'. Conan Doyle, of course, sent it straight back with his own note attached: 'I entirely agree.'

Conan Doyle was a resourceful man, however, and knew he could be more than just an unsuccessful physician. Throughout his life, he tried his hand at many career paths, becoming at various times an unsuccessful soldier (too fat to enlist), an unsuccessful MP (too unpopular to get elected), an unsuccessful ophthalmologist (literally never managed to attract a single patient to his practice), and an unsuccessful celebrity cricketer (his team, the – *sigh* – the Allahakbarries, was founded by J. M. Barrie, counted among its players Rudyard Kipling, H. G. Wells, P. G. Wodehouse, G. K. Chesterton, Jerome K. Jerome, A. A. Milne, Walter Raleigh, and a whole heap of other dudes responsible for half the Classics section in Waterstones, and was really, really bad at cricket.)

But medicine, the army, politics and sport's loss was literature's gain, and rather than having to spend his time treating Victorian eyeballs or shooting impertinent foreigners, Conan Doyle was able to devote himself instead to writing stories about an antisocial junkie with a bizarre name and an affinity for deductive reasoning. His name, for some reason, was Sherlock Holmes.

But we're not here to talk about that. We're going to concentrate on the other things Conan Doyle spent his time doing. Like trying to talk to ghosts.

Séances, mediums, spirit possession – Conan Doyle believed in it all. He's even the guy who came up with that whole 'Curse of Tutankhamun' thing we were all scared of as kids. In fact, he called spiritualism 'the most important thing in the world', and he spent the equivalent of millions of pounds trying to prove its veracity. He was a member of the Ghost Club and the Society for Psychical Research, and travelled the world to promote his belief in the paranormal – frequently falling out quite publicly with other notable figures of the day.

Take Houdini, for example – probably Conan Doyle's most famous nemesis in the spiritualist world. The two first met in 1920, both at the height of their respective celebrity, and while they started out as friends, brought together by a shared interest in investigating the supernatural, the relationship soon turned sour. Kind of ironically, given Conan Doyle's reputation as the architect of critical thinking and Houdini's as ... well, Houdini, the guy whose name is as tied to superhuman trickery as Holmes's is to deductive reasoning, it was Houdini who was the rational one of the pair – he clocked that the séances weren't legit after the pair attended one that claimed to be channelling his mum. She apparently wrote him a long letter in perfect English, signed with a cross, which was such an odd choice for a Jewish woman who only spoke Hungarian that Houdini figured it must have been fake. But Conan Doyle? He was taken in completely by various people claiming to be able to speak to the dead, over and over again.

Now, we shouldn't think he was a willing fool in this; he actually spent a reasonable amount of time and energy trying to debunk spiritualism before he was convinced, which honestly just makes it worse. He would frequently declare brazen frauds to be the real magical deal, including some that were specifically designed to show him how easy it was to fake those spiritualist feats he believed in so devoutly.

So, for example, there was William S. Marriott – or, when he was on stage, the magical Dr Wilmar. When he wasn't performing stage magic, he was acting as a sort of proto-Derren Brown or James Randi, going around duplicating the claims of so-called psychics and mediums and making damn sure people knew how he – and not some dead rando with too much spare time on their hands – did it. In December 1921, he invited Conan Doyle and three witnesses to a photo session in which he took a few snaps of the writer. Everyone attested there was no chicanery involved, it was a perfectly normal camera, and everything was above board and extremely not suspicious.

But when the photos were developed, there was a ghostly translucent figure behind Conan Doyle. It was nothing he hadn't seen before – the only difference was that, this time, the photographer wasn't pretending the images were kosher. They were published in the *Sunday Express*, along with witness statements as to the apparently normal photo and development process and an explanation from Marriott as to how fake the images were, which is to say, 100 per cent fake.

So Conan Doyle put out a statement of his own, so that everyone would know how totally Not Mad he was. 'Mr Marriott has clearly proved a point that a trained conjurer can, under close inspection of three critical pairs of eyes put a false impression upon a plate. We must unreservedly admit it,' he wrote, before claiming that, actually, he had known Marriott was a fake all along because he had the wrong hands to be a real medium.

'A conjurer has certain physical characteristics,' he wrote, including 'long, nervous artistic fingers'. *Real* spirit-talkers, he said, had hands that were 'short, thick and work stained', and so even if he might be taken in by Dopey, Doc or Sneezy, he wasn't going to be fooled by old lady-fingers Marriott over here.

Now, some people may consider such a reaction to be 'petty' or 'embarrassing', but it's actually remarkably sober compared to some of the other times he got caught out. The reason Houdini was such a sceptic towards the whole spiritualism schtick was largely because he'd been in Vaudeville basically his entire life, and he knew only too well how easy it was to fool a willing audience. When he tried to explain this to Conan Doyle, he was met with denial. And not just like 'well, sure you can replicate it, but these guys did it for real' kind of denial – Conan Doyle straight-up told Houdini he was lying about not being a magical being with powers granted to him from beyond the veil. Which, when you've spent your whole life putting in the mental and physical effort to perfect your stage act, isn't actually that nice a thing to hear.

The same was true of the 'Masked Medium', a woman seemingly able to conjure ghostly spirits to tell her the contents of a locked box filled with personal items brought by audience members, which certainly makes the afterlife sound like a pretty boring gig. But she got every single personal item right, in detail, and Conan Doyle was won over.

The problem was, the woman wasn't a psychic or a medium or a spirit channeler – she was just a tech nerd. That 'mask' – a veil, in fact, which the performer never took off – hid not just her face but a small wireless radio, with an assistant on the other end listing off the items in the box. The 'ghost' she had summoned had been an extra, waving a gauze sheet around in the dark. So when Conan Doyle heard that the act had come clean, and weren't psychics at all but psychic debunkers, he had quite the cognitive dissonance going on.

So what did he do? The same thing he had with Houdini: he told everyone that the Masked Medium *was* psychic, and that her explanation for how the trick had been done was a lie, and that even if it wasn't a lie that didn't mean it hadn't been real when *he* saw it. 'There is nothing to show that the first séance was not genuine,' he protested.

And yet none of this even comes *close* to Conan Doyle's greatest claim to credulous fame. That came in 1920, after he first heard about a series of photographs taken by two young girls – cousins, aged sixteen and nine, from the village of Cottingley, West Yorkshire.

The 'Cottingley fairies', as they're now known for fairly obvious reasons, is one of the most notorious hoaxes in history – if only because it's so unbelievably daft. It was essentially a photoshop job: the two girls, Elsie Wright and Frances Griffiths, had produced five photos that they said showed proof of fairies.

Yes, as in Tinkerbell.

It was most likely an innocent prank that got out of control. After all, doctoring photos of themselves to include things like fairies, or unicorns, or Transformers, or silly things like that – that's the kind of thing any kid might do. But, unfortunately, their mum moved in the same kind of circles as Conan Doyle, and that meant she was about as gullible as he was too. She sent the photos to a bunch of spiritualist magazines, saying they were the real deal, and, suddenly, the kids were famous.

But they likely wouldn't have become *so* famous, had the creator of the most popular literary character in history not taken up their cause. Conan Doyle took to the national press, writing a long article in *The Strand* magazine titled 'Fairies Photographed – An Epoch Making Event Described by A. Conan Doyle', in which he called the images 'the most astounding photographs ever published'.

'It seems to me that with fuller knowledge and with fresh means of vision, these people' – by which he means fairies, just to be clear – 'are destined to become just as solid and real as the Eskimos,' he wrote, Victorianly.

'These little folk who appear to be our neighbours, with only some small difference of vibration to separate us, will become familiar. The thought of them, even when unseen, will add charm to every brook and valley and give romantic interest to every country walk.'

Now to be fair to Conan Doyle, he did carry out what was, in his mind, a thorough investigation into the authenticity of the photos. Unfortunately, this being a hundred years ago, the definition of 'thorough investigation' was somewhat looser, and consisted mostly of Conan Doyle saying things like 'two working-class girls wouldn't be able to fool me!' and 'yes, OK, their house is full of paintings they've done of fairies but I just don't think they painted these ones, OK?'

Faced with such compelling evidence, it's no wonder Conan Doyle missed the more subtle clues that the photos were fake, such as the fact that the fairies were all obviously two-dimensional, and also fairly blatant copies of illustrations from a popular children's book of the time, or, lest we forget, mythical creatures. Even at the time, there were people pointing out how odd it was that the fairies all seemed to be dressed in the latest French fashions, and how despite being the creators of what at least one popular mystery novel writer claimed would 'mark an epoch in human thought', the girls didn't seem all that interested by the appearance of fairies in front of their face.

'For the true explanation of these fairy photographs what is wanted is not a knowledge of occult phenomena but a knowledge of children,' read *Truth*, a newspaper from Sydney, while the novelist Maurice Hewlett wrote that 'knowing children, and knowing that Sir Arthur Conan Doyle has legs, I decide that [the girls] have pulled one of them', which is basically just an old-timey way of saying 'you've been had, Conan Doyle'.

Conan Doyle was undaunted. He had, he pointed out, taken the photos to various technical experts in companies like Kodak and Ilford, and the majority had said they were real. This was a rather optimistic claim on his part, as the experts had mostly said things like 'these photos are fake' or 'well, they probably weren't faked in a lab, but that doesn't mean they're proof of fairies existing, you get that, right, Arthur?', but we should give him his dues here: some people were even more credulous. The novelist Henry de Vere Stacpoole, for example, said the photos were real because the girls just, you know, *looked* honest.

'Look at [Frances's] face. Look at [Elsie's] face,' he wrote. 'There is an extraordinary thing called Truth which has ten million faces and forms – it is God's currency and the cleverest coiner or forger can't imitate it.'

Of course, the trouble with loudly proclaiming to the world that you believe in fairies is that the world may well proclaim right back that you're an idiot, and that's exactly what happened to Conan Doyle. Gradually, the world – or, at least, that part of it that believed in fairies – quietly dropped the topic, and started regarding the man who gave them Sherlock Holmes as a bit, well, embarrassing and unhinged.

Not that he would have minded too much. He'd have been upset that so much of the world today rejects the idea that ghosts and fairies and spiritualists are among us, but he made it quite clear throughout his life that, if his reputation had to suffer for his belief in Tinkerbell and Smurfs, then so be it. And to illustrate just how far he'd come from 'elementary, my dear Watson', he explained himself thus:

> I have always held that people insist too much upon direct proof. What direct proof have we of most of the great facts of Science?

… Only the ignorant and inexperienced are in total opposition, and the humblest witness who has really sought the evidence has more weight than they.

In 1930, Conan Doyle finally got the opportunity to penetrate the veil between the living and the dead for himself, which is to say: he died. Despite his insistence that spirits can pass into our world and effect change or send messages, he appears to have slept through Elsie and Frances admitting, more than half a century later, that the photos were indeed fake.

'It was just Elsie and I having a bit of fun,' Frances told the BBC in a 1985 interview. But just imagine: you're sixteen, and messing around with your cousin, and the world's most famous writer of the world's most famous incredibly clever guy comes along and publishes your photos for the whole world to see and writes a long tract about how your fun camera experiment is proof positive that fairies exist and the whole world rests on these photos … well, what are you going to do?'

'Two village kids and a brilliant man like Conan Doyle – well, we could only keep quiet,' explained Elsie.

'I can't understand to this day why they were taken in,' said Frances. 'People often say to me, "Don't you feel ashamed that you have made all these poor people look like fools? They believed in you." But I do not, because they wanted to believe.'

Of course, perhaps it wouldn't have made any difference if Conan Doyle had found out the photos were a fake. After all, he'd probably say, just because *those* ones were frauds, doesn't mean fairies don't exist. The girls probably had the wrong shaped hands to take *real* photos of fairies.

But you would have to think he'd be disappointed. When he looked at those photos, he didn't just see two girls sitting next to what are clearly paper cut-outs of fairies – he saw the dawning of a new age.

'When Columbus knelt in prayer upon the edge of America, what prophetic eye saw all that a new continent might do to affect the destinies of the world?' he wrote in *The Strand*. 'We also seem to be on the edge of a new continent, separated not by oceans but by subtle and surmountable psychic conditions. I look at the prospect with awe ... there is a guiding hand in the affairs of man, and we can but trust and follow.'

22.

Thomas Edison's Lesser-Known Invention: Dial-a-Ghost

By the time Thomas Alva Edison died in 1931, he had patented more new inventions than anybody else in history. He achieved this in much the same way as the pharaohs of Egypt built the pyramids, which is to say, he got a bunch of other people to do it, and then slapped his own face on the finished product.

That's not totally a bad thing – Edison-heads these days often credit him with 'inventing' the concept of independent industrial research labs more than anything else, and that alone has given us ... well, pretty much every modern convenience you have at your disposal right now, probably. But the idea you likely have in your mind of a brilliant loner being hit by bolt after bolt of divine inspiration is pretty much a fabrication: as the man himself once said, 'I never had an idea in my life.'

In fact, most of even his most iconic creations were the result of collaboration with dozens, hundreds, or even thousands of the employees working at his Menlo Park laboratory – his 'invention factory', as he affectionately called it. I mean, the guy didn't even *really* invent the lightbulb, and that's the one thing he's best known for inventing.

Actually, the invention of the lightbulb might be the best way to understand Edison as a whole. It wasn't one man's epiphany, but the culmination of decades of various scientific experiments from the worldwide crank community. Depending on how you judge it, the original inventor was either: Ebenezer Kinnersley, who in 1761 demonstrated that a piece of wire could be heated into incandescence; Humphry Davy, whose 1802 'electric arc lamp' consisted of an electrified strip of platinum and was used for decades in the street lights of various European cities; Frederick de Moleyns, who was granted the first-ever patent for an incandescent lamp in 1841; Joseph Swan, who made the first carbonised filament bulbs in 1850; Moses G. Farmer, whom Edison asked for advice on lightbulb-making after seeing Farmer's bulbs in use in 1859; or, failing that, any of the dozens of other inventors and scientists who had created a way to light a bulb using electricity before Edison came along.

Edison's claim to have 'invented' the lightbulb is shaky at best, is the point. His genius wasn't in the invention itself, but the process – he was the one who assembled the shop-floor teams and got them fine-tuning the established process, by testing material after material for years on end in search of the best filament. Then, after they got the general idea down, Edison patented the design and founded the Edison Electric Light Company well before they perfected it – his patent is remarkably vague on details outside of 'carbon filament in glass bulb – T. A. Edison'.

Perhaps more than anything, though, the true story of the lightbulb highlights one of the most valuable talents Edison had at his disposal: he was, in modern terms, incredibly media-savvy.

Let me explain: why do you think we associate Edison with the lightbulb? We've already established he didn't exactly come up with the idea himself: if electrical illumination was really that important, this book would have a chapter on Ebenezer Kinnersley, whoever he was. Neither was it his first patent – that had been an electronic vote-recorder he invented a decade earlier to remove the need for tallying by hand.*

Was it the invention that made him famous? No: that was the phonograph, which two years earlier had turned Edison from freelance to celebrity inventor. Even more impressive was the fact that he had invented it despite being almost entirely deaf: 'I haven't heard a bird sing since I was twelve', as he once said. He got around this, when he needed to, by chomping into the wood of a piano or phonograph and 'listening' to the sound through his bones, a habit which sounds made up for comedic value but I assure you is true.†

OK, so maybe it was because the lightbulb was so transformative – after all, he may not have made the *first* one, but he made the one we actually *used* for a hundred years, right?

Well, again, not really – Edison's bulbs worked by electrifying carbonised bamboo, which was cheap and effective enough to mass produce, and better than buying 200 candles every month, but not *great*. They were pretty short-lived: an average lifespan of around 100 hours meant you'd probably be burning through one or two every month. One or two at *least*, I should say, because that carbon filament meant that they were both very fragile and prone to dimming over time due to chemical reactions with the oxygen inside the bulb.

---

* This invention, too, is a good lens through which to view Edison, because it was the one that – for want of a better word – radicalised him. When he dreamed up the machine, he thought his fortune was made: his invention would save hours of public time and remove the need for tedious labour! But, in fact, it was a crushing failure: no legislators were interested in a device that would count votes so quickly that there would be no time to lobby for extra support while the ayes and nays were being submitted.

From that point on, Edison vowed never to invent something that he couldn't already see a market for. Now, to be fair, it is difficult to invent anything at all with an empty stomach, but this decision may be one reason his modern reputation is, let's say, less romantic than contemporaries like Tesla.

† That meant two things: one, tooth marks in all his phonographs, and two, his opinion that music by legends like Mozart or Rachmaninoff was utterly worthless.

It didn't take long before various tinkerers started improving and updating the design, and, by the mid-1910s, the vast majority of lightbulbs housed tungsten, not carbon, filaments.

So what was it that fixed lightbulbs so firmly in our minds in relation to Thomas Edison? Quite simply, it was the guy's outstanding PR game. He was photographed next to the invention so often that it kind of became part of his persona – and, perhaps by extension, the symbol of invention itself. He combined the bulbs with his newly-patented 'direct current' system of electricity distribution, and lit first the entirety of Menlo Park and then part of New York City – members of the public would literally congregate in the streets to marvel at the lights that could burn as bright as sixteen candles and be dismissed at will, which just goes to show what kind of thing passed for 'entertainment' back before the invention of reality TV.

Basically, you can think of Edison as a sort of proto-Elon Musk. The papers loved talking about his successes and the famous 'Edison Test' that he set potential new hires*. They loved making fun of him too – my personal favourite example is the cartoon of well-to-do Victorian ladies and gentlemen flying around high society in loop-de-loops thanks to 'Edison's Anti-Gravitation Underclothing', which sounds even worse than hoverboards in terms of unrealised dreams. They published his opinion on pretty much anything – including things like the existence of God or the economic effects of quantitative easing – and at least one newspaper even published a whole sci-fi novel about him. And perhaps that's why he, like so many celebrity geniuses before and after him, fell into the most dangerous trap of all: believing your own hype.

---

* Imagine you're applying for a job as an engineer, you walk into the interview room, and instead of being asked 'When was the last time you worked in a team, and what obstacles did you overcome?', you were given a pop quiz featuring questions like 'Where in Germany do toys come from?' and 'What voltage do trams run on?' The 'Edison Test' was at least 146 questions long, and loudly hated by those who failed it – a group that included such luminaries as Albert Einstein and Edison's own son.

Thomas Edison's Lesser-Known Invention: Dial-a-Ghost

See, for somebody who relied so much on collaboration, Edison sure was bad at teamwork. Inventions that came from the community at Menlo Park were 'mine' rather than 'ours', and inventions that came from anywhere else were worthless. He hated people tinkering with his designs so much that it actively hampered his ability to make money: for instance, when he learned that people were playing records on his phonograph too fast – *on purpose*, mind you – instead of leaning into it, he directed his workers to modify the machines to make it impossible. He refused to make his records compatible with other companies' phonographs, and reserved the right to veto any record he didn't personally enjoy – a right which he exercised, calling the most popular music of the day 'miserable', for 'the nuts', and reminiscent of 'the dying moan of dead animals'. Famously, he went on the warpath against Tesla and Westinghouse to promote his own direct current system over their alternating option by electrocuting dogs and criminals and making elephant snuff films,* a period of history that's become known as the 'War of the Currents' and not, for some reason, 'the time Thomas Edison murdered several dogs and an elephant'.

Edison was a man unable to conceive of failure. Had he been alive a hundred years later, he'd be one of those tiresome bosses who keep telling you that the Chinese characters for 'problem' and 'opportunity' are the same thing – they aren't, as approximately 1.5 billion Chinese speakers will tell you – but, instead, he reputedly said, 'I have not failed. I've just found 10,000 ways that won't work.'

---

* 1903's *Electrocuting an Elephant*, which shows the – for want of a better word – execution of Topsy the elephant. Edison's selling point for why direct current was better than the alternating kind hinged on the latter's higher voltage, which he said made it more dangerous – 'Westinghouse will kill a customer within six months after he puts in a system of any size,' he wrote in 1886.

To prove this, he suggested his competitor's electricity system as a means of execution for both elephant and human criminals, and when there were no criminals to electrocute, he used dogs instead. He even tried to make people say 'Westinghouse' instead of 'electrocute', but it never caught on.

Well, no offence to Edison, but here's the thing: he failed a bunch. And not just because he lost the War of the Currents* and thought radio and movies were vulgar fads; this guy had some real stinkers in his repertoire. Take, for example, his concrete furniture – 'for use in … concrete houses', the reporters of the day explained – or his talking dolls that were so creepy they had to be taken off the market less than a month after being released. We think of him today as, if nothing else, a genius with a string of world-changing inventions to his name, but, really, Edison was just playing the averages – and nothing encapsulates that more than perhaps his weirdest contraption of all: the Spirit Phone.

'I have been at work for some time, building an apparatus to see if it is possible for personalities which have left this earth to communicate with us,' he told *The American Magazine* in a 1920 interview.

This coming invention would not use 'any occult, mystifying, mysterious, or weird means, employed by so-called "mediums", but by scientific methods,' he assured the public. 'I am engaged in the construction of one such apparatus now, and I hope to be able to finish it before very many months pass.'

Before you could say 'how's the reception in Hades these days', magazines across the world were publishing articles and cartoons lampooning the inventor who could talk to the dead. But to Edison, it made perfect sense: with his usual self-belief, he told the magazine that he knew 'with absolute positiveness that some of our most generally accepted notions on the subject [of life after death] are utterly untenable and ridiculous'. In fact, he said, all living things are made from 'myriads and myriads of infinitesimally small individuals, each in itself a unit of life'. These individuals were immortal, lived in some kind of vague hierarchy – to 'account for the fact that certain men and women have greater intellectuality, greater abilities, greater powers than others,' he explained – and worked together in what he called 'swarms' to generate intellect and personalities.

---

* Direct current was only useful over very short ranges. You can tell he lost the war by the lack of power plants surrounding your house.

Contacting the dead, then, wasn't a question of séance but science – all you need to do is construct a way of isolating these 'swarms'.

'If the units of life which compose an individual's memory hold together after that individual's "death", is it not within range of possibility to say the least, that these memory swarms could retain the powers they formerly possessed, and thus retain what we call the individual's personality after "dissolution of the body?"' Edison asked the readers of *The American Magazine*.

'If so, then that individual's memory, or personality, ought to be able to function as before ... I am hopeful that by providing the right kind of instrument, to be operated by this personality, we can receive intelligent messages from it in its changed habitation, or environment,' he said.

It sounds absurd, I know – which is why, for about a century, biographers mostly thought it was a hoax. But, for the time, it wasn't such a crazy idea: despite the best afforts of scientists and debunkers like Houdini, spiritualism – the belief that the dead were able to communicate with the living and even affect events in the real world – had been steadily gaining popularity throughout the second half of the nineteenth century.

It kind of makes sense as well: the whole century, people had been bombarded with inventions that opened up brand-new possibilities, and a bunch of them had come from Edison himself. I mean, sure, it might seem obvious to *us* that you can't talk to the dead, but it probably seemed obvious to people two hundred years ago that you can't listen to the opera if you're at home – then Edison invented the phonograph.

And at the time Edison was out there talking about his spirit phone, spiritualism was all the rage. By 1920, two things had happened that had left the world reeling: the First World War, and the 1918 pandemic. Parents had lost their sons to foreign trenches – quite often they never even saw their child's body again – and then, when the dust settled, along came a deadly flu to take out the survivors. In total, the second half of the 1910s saw the loss of around 70 million people – around one in twenty of everybody alive – and most of them were young and healthy, with whole lives ahead of them.

In the wake of so much death, people scrambled for any sense of control – and with spiritualism, they were promised their sons and daughters were still with them, and happy. Books on spiritualism sold out over and over; in America, Ouija boards started selling out, and in the UK, spiritualism became so popular that it briefly gave the Anglican Church a run for its money.

So despite his biographers' attempts to cover it all over with a 'ha ha, that was probably a joke, he didn't really think you could phone up ghosts', Thomas 'Will Invent for Money' Edison almost certainly was serious about this suggestion. And in 2015, any lingering doubts about whether he really meant it or not were put to bed when a French radio presenter named Philippe Baudouin happened to find an old copy of Edison's memoirs in a second-hand bookstore.

The 'Diary and Sundry Observations' of Edison was essentially just a bunch of his writings and speeches that had been collected up and sold seventeen years after he died, which, to be fair, sounds like what he would have wanted. Whoever organised the book certainly had a lot to choose from, since Edison left approximately five million pages of notes, letters, diaries and the like after he died,* but for some reason – and let's face it, it was probably embarrassment – they neglected to include any mention of the spirit phone.

At least, in the English version. In French, the book had one extra chapter: Edison's designs for the spirit phone. Sketches showed that he was basing the idea off his phonograph designs – this was a standard way of working for him, a technique that modern biographers have called his 'invention by analogy'. His idea seems to have been to ramp up the sensitivity of the phonograph enough that it would be able to pick up the vibrations of the swarm. He even made

---

* The Edison Project, which aims to publish and preserve these documents, was set up in 1978. It's not finished yet.

a pact with one of his engineers, William Walter Dinwiddie, that whichever of them died first would send the other a message from beyond.

Even if his chronicler was ashamed of these occult beliefs, Edison himself had no such qualms. Ever his own biggest hype man, he told *The American Magazine* that 'the apparatus ... should provide a channel for the inflow of knowledge from the unknown world – a form of existence different from that of this life – we may be brought an important step nearer the fountainhead of all knowledge, nearer the intelligence which directs it all.'

To be fair, if it had worked, it probably would have beaten even the lightbulb in terms of world-changing inventions. Luckily, though, ghosts can't use phones,* so the plans came to nothing. Nobody ever found a prototype, and it was generally assumed that, like so many other ideas he had at some point been obsessed with, Edison had simply dropped it and moved on to the next distraction.

Well, kind of. Two years after he died, in 1933, the magazine *Modern Mechanix* published an article claiming that Edison had convened a group of eminent – though unnamed – scientists several years previously to test a prototype spirit phone.

'Edison set up a photo-electric cell,' the story explained. 'A tiny pencil of light, coming from a powerful lamp, bored through the darkness and struck the active surface of this cell, where it was transformed instantly into a feeble electric current. Any object, no matter how thin, transparent, or small, would cause a registration on the cell if it cut through the beam.'

The group spent hours watching the machine for any sign of communication, the article said, but to no avail. If the phone was

---

* I say luckily because, well, imagine if they could. An eternity of just watching us all like the world's most boring soap opera? They'd never stop ringing us up and complaining. (In fact, a French cartoon from the time showed just that: a depressed husband being henpecked by his dead mother-in-law.)

ringing, nobody was picking up on the other side. And that, perhaps, is the strongest evidence of all that the spirit phone was a flop. After all, Edison was infamous in his day for being excessively litigious – if the device had worked, there's no doubt his spirit would have got on the phone to *Modern Mechanix* and threatened legal action.

23.

Real-Life Supervillain Nikola Tesla Takes the Term 'Pigeon Fancying' a Bit Too Literally

The more you learn about Nikola Tesla, the less you believe he could possibly have been a real person.

This was a man who, in 1926 – that's *nineteen bloody twenty-six* – was talking about being 'able to communicate with one another instantly, irrespective of distance … we shall see and hear one another as perfectly as though we were face to face, despite intervening distances of thousands of miles; and the instruments through which we shall be able to do this will be amazingly simple compared with our present telephone. A man will be able to carry one in his vest pocket.' Sound familiar? Yeah, hand over the iPhone patent, Jobs – Tesla beat you by eighty years.

This was a man who was demonstrating wireless drone technology in *1898*, a time when your average person's reaction to seeing a remote-controlled boat zipping across the water was to assume there must be a tiny monkey hiding in the captain hole and swizzing the little spikey wheel about to control it.* This isn't hyperbole by the way – this is literally what some in the crowd accused Tesla of doing.†

This was a man who predicted that although 'today the most civilized countries of the world spend a maximum of their income on war and a minimum on education. The twenty-first century will reverse this order ... The newspapers of the twenty-first century will give a mere "stick" in the back pages to accounts of crime or political controversies, but will headline on the front pages the proclamation of a new scientific hypothesis.'

OK, so two out of three ain't bad. But you get the picture: Nikola Tesla may be remembered today as an engineer and inventor, now most famous for inventing the alternating current motor, but he really might as well have been some kind of time-travelling wizard.

He looked the part too: 6 foot 2 inches tall, extremely slender, with striking eyes and meticulous clothes and hair. He spoke eight languages and slept only two hours per night. He is the only physicist to have ever been played by David Bowie in a movie, and probably the only person in existence too eccentric for even the Thin White Duke himself to do them justice.

Tesla's fascinating life and arguably fascinating-er inventions could – and do – fill multiple books, and if you're interested in learning more about the nature of his genius, I really suggest you close this book and read one of them instead, because what we're going to talk about here is less Tesla the visionary engineer and more Tesla the aging serial hotel squatter and real-life Bond villain.

Depending on how you look at it, Nikola Tesla either spent the entire second half of his life in opulent luxury or literally homeless.

---

* I don't know much about boats, I hope it doesn't show.
† The less rational onlookers thought it was magic, or telepathy.

He returned from his Experimental Station in Colorado Springs* to his adopted hometown of New York in 1900 and set himself up at the Waldorf Astoria, where he lived for nearly twenty years entirely on credit and the goodwill of his famous friends. After running up a bill of over $20,000 – that's more than half a million of today's dollars – he left the Waldorf and moved to another luxury Manhattan hotel, the St Regis, where he lived for a year before getting evicted for once again not paying his bill. From then on, he moved from hotel to hotel, getting kicked out of some of New York's most iconic establishments; eventually, in 1934, he moved to the New Yorker, where he was able to stay for nine years by virtue of somebody else paying his bill for him.

While he wasn't running from hotel bailiffs shaking their fists at him and yelling 'Teslaaaa!' like the straight man in a Hanna-Barbera cartoon, Tesla split most of the later period of his life between his three loves: pigeons, doomsday devices, and elaborate and kind of baffling birthday parties.

The parties started when he was seventy-five. A young science journalist and Tesla-fan named Kenneth M. Swezey threw him a birthday celebration, including a spread inside and cover of *Time* magazine, and praise and well-wishes from, among others, Albert Einstein, which I'm betting is more than you or I ever got for a birthday. Nikola had such a blast that he decided to make the party an annual tradition, and while that sounds annoying as hell for everyone around him, I can't really say I wouldn't do the same if

---

* It is here that the writers of the book and film *The Prestige* would have you believe he was helping a couple of English magicians clone themselves while looking suspiciously like Ziggy Stardust. The truth, as ever seems to be the case with Tesla, is even weirder: he had been making headlines for communicating with aliens. Although it's *fairly* unlikely that he really had been receiving messages from Mars (although, if anybody were to do it, it would have been him, let's face it), to this day we don't know what the signals he heard actually were. The most probable explanation, though, is that he was accidentally picking up experimental transmissions from rival engineers who were also trying to master wireless communication at the time.

I were a world-famous eccentric genius living rent-free in gilded-age Manhattan.

Each year he would throw an elaborate banquet and invite the press to listen to him opine on current events and science, making increasingly wild and confusing claims, like 'Einstein is an idiot, actually'* and '[I have perfected] a new small and compact apparatus by which energy in considerable amounts can now be flashed through interstellar space to any distance without the slightest dispersion ... for means of communication with other worlds'.†

Not content with potentially triggering an alien invasion, he also devoted a bunch of time to developing various ways to destroy all life on Earth by himself. He claimed he had a pocket-sized device that allowed him to, should the urge take him, destroy the Empire State Building. He espoused eugenics.‡ And then, of course, there was Teleforce.

The press at the time called Teleforce a 'death ray' or, somehow even more chillingly, a 'peace ray'. Nikola was aghast at this description: in 1937, he explained, presumably to a chorus of sighs of relief, that 'I want to state explicitly that this invention of mine does not contemplate the use of any so-called "death rays".'

'Rays are not applicable,' he continued, presumably to a chorus of people immediately regretting their sigh of relief, 'because they cannot be produced in requisite quantities and diminish rapidly in intensity with distance ... My apparatus projects particles which ... convey to a small area at a great distance trillions of times more energy than is possible with rays of any kind.'

Got it? He's not the crazed inventor of a terrifying death ray. He's the crazed inventor of a terrifying death *beam*.

---

* Not an actual quote.
† This one is an actual quote.
‡ Before you jump in with 'Yes, but everybody did at that point' – yes, it's true that eugenics was less of a dirty word before the Nazis showed the world the inevitable conclusion of the idea, but Tesla was talking at a point when many US states were openly performing forced sterilisation on so-called 'undesirables', and he knew about it, *and he thought that wasn't going far enough.*

'All the energy of New York City (approximately two million horsepower) transformed into rays and projected twenty miles, could not kill a human being, because, according to a well-known law of physics, it would disperse to such an extent as to be ineffectual ... [With my invention] many thousands of horsepower can ... be transmitted by a stream thinner than a hair, so that nothing can resist,' he added, just to really make it clear how much time he hadn't spent thinking about murdering people with death beams, honestly.

Luckily, almost none of these theories or inventions made it past the 'trust me, guys, I've definitely done it, no you can't see the prototype, it's in my other workshop, in Canada, honest' stage of production. This may well be in part because of Tesla's other great passion in life: pigeons. He was famous in New York for walking every day to the park to feed the pigeons; he would nurse sick or injured birds back to health in his hotel rooms* and was known to have asked hotel chefs to prepare special dishes to feed his feathered friends. He truly loved them.

And when I say 'loved', I mean he literally claimed to have fallen in love with a pigeon.

'I loved that pigeon as a man loves a woman, and she loved me,' he said of, just to be absolutely clear here, a pigeon. 'As long as I had her, there was a purpose to my life.'

Unfortunately, though unsurprisingly for anybody who has ever looked up the average lifespan of a pigeon, the relationship ended tragically. In 1922, he said, his beloved pigeon flew to his hotel room to tell him she was dying. In her eyes, he saw 'two powerful beams of light ... a powerful, dazzling, blinding light, a light more intense than I had ever produced by the most powerful lamps in my laboratory.'

Her boyfriend's lamp game suitably humbled, the pigeon died in his arms. Tesla later said that, at that moment, he knew his life's work was complete, which is a sweet but arguably overdramatic reaction

---

* This was, perhaps unsurprisingly, another reason he would get evicted so much.

to a pigeon death. Tesla would eventually join her in the Great Big Dovecote in the Sky in 1943, at the age of eighty-six, when he died alone in his hotel room, a penniless legend.

But this story does have an epilogue, because even death couldn't stop Nikola Tesla from ripping off luxury hotels. Unable to pay his overdue bill at the Hotel Governor Clinton, he offered to give them as collateral a box containing a working model of his death beam – on the strict understanding that they were never to open it. They fearfully accepted the condition, presumably figuring that something a reclusive visionary inventor terms his 'death beam' probably isn't something you want to be messing about with. Until, that is, he died, at which point they opened the box to see what kind of world-changing technology they were sitting on.

They found nothing but a few harmless electrical components. Which is just as well, really, because if you're running the kind of establishment that accepts cash, credit or Death Beams with Which to Smite Your Enemies, then it's … probably for the best that you're not given access to a fucking death beam.

24.

Marie Curie Defies All the Odds to Accidentally Poison Both Herself and Thousands of Strangers

Every branch of science has its own safety protocols. In engineering, for instance, they set a lot of store by proper paperwork. Archaeologists have to learn which discoveries are the longed-for solutions to centuries-old mysteries and which are last week's crisp packet. And in chemistry, the number one rule is simple: please, for the love of God, do NOT mess about with the random glowy substance you've just discovered in a poorly ventilated and leaky shed.

It's not known when this rule first made it into the standard lab protocol handbook, but it must have been after 1898, when a young scientist named Marie Curie was busy doing exactly that.

It wasn't obvious, when Maria Salomea Skłodowska entered the world in 1867, that she would grow up to leave the legacy she did. She was born in Warsaw, which we think of now as a city in Poland, and many people at the time thought of as a city in Poland, but the Emperor of Russia, Tsar Alexander III, was very sure was a city in Russia. Luckily, the government in Warsaw was extraordinarily liberal for the time, and voicing one's disagreement with the Tsar's opinion was not only allowed, but taken as agreement to have all your earthly possessions and potentially your life taken from you by force. Ma and Pa Skłodowska,* being of the view that a country full of Polish people speaking Polish and saying things like 'ach, Stanisławie, cóż to za piękny dzień na bycie Polakiem'† was probably Poland and not Russia, had taken part in some of the fairly regular uprisings aimed at restoring Polish independence. As a result their five children were born into a family whose circumstances had been reduced from 'a level of comfort befitting the household of two successful and well-respected headteachers' to 'your sister Zofia died because we had to rent her room to a typhus patient'.

And poverty wasn't the only hurdle the young Maria would have to overcome. She was also (as you may have guessed) a girl – which made getting a university education illegal in Russian-ruled Poland. So, while their brother Józef went off to the University of Warsaw to study medicine, the Skłodowska sisters were forced to continue their educations through more clandestine means.

Here's the thing about Poland: everyone thinks they want it, but nobody can handle it once they've got it. Just ask the Swedes. Or the Prussians. Or the Austrians. Or the Germans. Or the USSR. Or any of the other forces throughout history who have tried to invade Poland and suddenly found themselves having to deal with a well-organised and highly motivated resistance movement. And it was no different in Maria's time: no matter what tools of oppression Imperial Russia tried to use on the Poles – be it banning the language, clamping down on education, deploying secret police or doling out draconian punishments to agitators – there was an admirably stubborn refusal throughout the country to submit.

---

* Technically Ma Skłodowska and Pa Skłodowski, due to how Polish names work.

† Which translates to something along the lines of: 'Gee Stanislav, it sure is a great day to be Polish!'

Maria was a smidge too young to have been a tax-paying* citizen of the underground Polish National Government of the mid-1860s, but even in the slightly subdued Warsaw of the 1880s, there were still opportunities for an intelligent Polish-speaking young woman in search of an education. So having completed their high-school educations, Maria and her sisters 'enrolled' in Warsaw's soon-to-be-infamous 'Flying University': an underground organisation dedicated to educating Polish youth with forbidden and dangerous knowledge, like Polish History, or Polish Literature, or Literally Anything Taught to a Woman. The lectures took place in various locations throughout Warsaw, changing regularly to avoid detection by the Russian authorities (this is why it was a 'Flying' University); there was even a secret library for all your illegal research needs. All in all it was an incredibly badass way to study maths and physics, which is what Maria did while nominally working as a governess and teacher.

Unfortunately, there aren't many graduate-level positions open for somebody whose alma mater doesn't officially exist, and the Skłodowska sisters were forced to leave their beloved Poland and study for a real, non-flying, degree. Being as poor as they were, Maria and her sister Bronisława made a deal: Bronya would go and study medicine in Paris for two years while Maria supported her financially from Warsaw, and then Maria would follow later and study science and mathematics while supported by Bronya. And that's science and maths as two separate degrees, by the way, not one joint degree. Maria was what you might call an 'overachiever'.

It was in Paris that young Maria Skłodowska would become Marie Curie – both legally and symbolically. Legally, because she met a charming young physicist named Pierre and the pair fell madly in

---

* Yes, really! The totally banned Polish National Government of 1863–4 was a big problem for the Russian regime, and it took its name very seriously: it had five permanent ministries, security and intelligence departments, 'ambassadors' in foreign capitals, and a capable and efficient Treasury funded by voluntary taxes from patriotic citizens (as well as involuntary ones from the less patriotic).

love; they married in 1895 – Marie, never one for unnecessary luxuries, wore the same dress for her wedding as she did in the lab – and she became Madame Curie. And symbolically, because it was in Paris that she would carry out the work that would make her the first and to this day only person *ever* to win two Nobel Prizes in two separate sciences.

It all started one overcast day in March 1896, in the office of a physicist and amateur photographer called Henri Becquerel. The previous November had seen the accidental discovery of a mysterious new type of radiation; they were known by some as 'Röntgen rays', after the physicist who had found them, but Wilhelm Röntgen himself preferred to call them 'X-rays'. Either way, the scientific community was obsessed, and Becquerel was no exception. He immediately started looking for a connection between these new X-rays and the subject of his own work: a rare but fairly boring element, mostly used to tint glass, called uranium.

Of course, now we know that uranium is neither rare nor boring: it can be found naturally throughout the world in soil and water, and unnaturally in various facilities where nuclear weapons are stored under adequate-to-terrifyingly-lax security measures. And in March 1896, it could very specifically be found half-forgotten on top of an unexposed photographic plate in the desk drawer of Henri Becquerel.

Now, this drawer hadn't been opened for days. Yet when Becquerel opened it, he found the image of the uranium crystals burned clearly into the plates. Clearly, he had stumbled upon something incredibly important, but he wasn't quite sure exactly what – some kind of invisible rays, possibly similar to X-rays, was his guess. The real answer, which we know now but for which Becquerel didn't even have a word at the time, was radioactivity – and it was Marie Curie who figured it out, confirmed it, and even came up with the words to describe it.

Unfortunately, what she didn't do was be in any way careful at all around it. In 1898, Marie and Pierre announced the existence of two new radioactive elements: polonium (named in honour of Poland) and radium (named after, uh, rays). Now, radium is highly, highly radioactive, and can cause health problems in a matter of hours. We know this because Pierre made notes about how his skin became burned and lesioned when he *strapped a vial of the stuff to his arm for ten hours*. Marie would keep vials of radium and polonium in her desk drawers; she would even walk around with them in her pockets. At home, she kept a sample of radium next to her bed to use as a nightlight.

'It was really a lovely sight,' she wrote in her autobiography. 'The glowing tubes looked like faint, fairy lights.'

Meanwhile, Pierre would keep a sample of radium in his pocket in case he happened to meet someone who wanted to see its warm and apparently not-suspicious-at-all glow. It was kind of like the romantic comedy *You've Got Mail*, except instead of book sales they were competing over who could expose themselves to the highest levels of radiation.*

'But that's not fair,' you might be saying right now. 'They didn't know radioactivity was poisonous – nobody had ever seen it before!' Well, first of all, I direct you to the sentence a few lines above where they were *strapping samples of radium to themselves and watching it burn their skin off in real-time*, and second of all, look: I've not won a Nobel Prize in physics, but if I came across a warm glowing substance that neither I nor anybody else had ever seen before, you know what I wouldn't do? Keep it on and near my person for days on end as a fun novelty item, that's what.

Marie and Pierre didn't even make the connection when their fingers (or as they thought of them, the handiest radium containers) became cracked and scarred. Marie would spend whole days stirring cauldrons full of boiling uranium ore – she personally processed literally tons of the stuff by the time her discoveries made it to professional review.

And they were *her* discoveries, to be clear: sure, finding polonium and radium had been joint efforts, but the radioactivity of uranium,

---

* So maybe it was more like *Flubber*.

and later thorium, came from Marie alone. And she made damn sure everyone knew it, too, even specifically stating twice in her biography of Pierre that the work was unambiguously hers. I know that sounds slightly petty, but it's really a good thing she was willing and able to stand up for herself like that, because the everyday misogyny she faced nearly denied her a Nobel Prize: the 1903 prize in physics was originally to be awarded to Pierre Curie and Henri Becquerel for the discovery of radioactivity – you know, despite Marie doing if not *all* the work, then at least more work than 'accidentally leaving some poison metal in a drawer'. Thankfully, Pierre, apparently a man ahead of his time, complained about Marie's lack of recognition, and her name was added, thus making her the first woman to ever win a Nobel.

But after years of working with these radioactive materials, both Curies complained of constant pain and fatigue, and eventually Marie died, aged sixty-six, of aplastic anaemia caused by radiation poisoning.* To this day, over a century later, her notebooks are so incredibly radioactive that they must be kept in lead-lined boxes; if you want to view her personal possessions at the Bibliothèque Nationale in France, you have to wear protective clothing and sign a liability waiver.

While it's undoubtedly sad that such an inspiring woman was killed by her own discovery, she definitely wasn't the worst case of tragic radioactivity-related ignorance. In the decades that followed her work, the world fell in love with radium, the amazing glowing energy source that, in the right hands, could literally cure cancer. And because this was the olden days, and capitalism is what it is, people started putting it in just about everything: face creams, soap, toothpaste, lipstick, watches, razor blades, suppositories and even condoms.

It wouldn't be until the 1930s that crackdowns would begin on this abundance of radioactivity on the high street, and it was a

---

* Pierre, by the way, managed to foil the cruel gods of irony by absent-mindedly walking into the street and getting run over by a cart in 1906.

particularly gruesome case that prompted it. Eben Byers, a socialite and athlete, was famous for drinking up to three bottles per day of a popular 'energy drink' called Radithor – he said that it gave him a 'toned-up feeling'. Unfortunately, what it was actually doing was poisoning his entire body with cancers that disintegrated his bones, caused his teeth to fall out and holes to form in his skull, and left abscesses on his brain.

'A more gruesome experience ... would be hard to imagine,' a federal attorney reported while investigating Radithor. 'We went up to Southampton where Byers had a magnificent home. There we discovered him in a condition which beggars description.'

When he died, Byers's bones were found to contain 36 micrograms of radium – nearly four times a fatal dose. He was buried in a lead-lined coffin. As the *Wall Street Journal* put it when they reported his death in 1932: 'The radium water worked fine until his jaw came off.'

25.

Albert Einstein: Public Nuisance, Love Rat

Einstein is most famous today as, well, *Einstein* – the guy who basically rewrote the entire way we understand the universe. Even in a book like this, he stands out: he's like the Hoover or Biro of geniuses, in that his name has basically become a synonym for what he was.

He won a ton of awards for his work in relativity and quantum mechanics, including the highest honour of the German Physical Society, the Max Planck Medal – which must have been a bit weird for Einstein, actually, since Max Planck himself was a good friend of his in real life – and Columbia University's Barnard Medal for Meritorious Service to Science; he was *Time*'s Person of the Century, and he, of course, won the Nobel Prize in Physics in 1929. He's even appeared in multiple *Animaniacs* cartoons, and those gigs are tough to get.

But amongst all these academic accolades is something that stands out as truly impressive: Einstein's uncanny ability to make literally dozens of complete strangers save his life, over and over again, purely through his own stupidity, and never once even *look* like he might intend on learning anything from it.

Albert Einstein: Public Nuisance, Love Rat • 243

There's a very famous photo of Albert Einstein – well, OK, there are loads of very famous photos of Albert Einstein, but this one in particular shows him sitting on a large rock, posing coquettishly in a polo shirt and blue shorts with a pair of women's sandals on his feet. The glorious blue ocean stretches out behind him as he talks to a local man who has, because it was the 1930s and that's apparently the kind of thing people did back then, decided to come to the beach in a full suit.

The year was 1939, and Einstein was on holiday. He had rented a cottage in the sleepy coastal hamlet of Cutchogue, Long Island, and he had just one thing on his mind: sailing.

Einstein *loved* sailing. He sailed in Switzerland as a teenager, and he sailed in New York as a sexagenarian. His beloved coastal cruiser, a present from his friends that in typical jocular fashion he called his 'thick little yacht', had been seized by the Nazis some years earlier, and he now spent his leisure time in her replacement, *Tinef* – a Yiddish word meaning, approximately, 'little piece of junk'. Relaxing on a boat, he said, he could be oblivious to the world.

There was just one problem: he was terrible at it.

'You had thirty people around here who'd tell you they rescued Einstein when he capsized, and towed him and his boat in,' recalled Robert Rothman, the then-twelve-year-old son of Mr Wears-Formal-Attire-at-the-Beach.

'We kids who were growing up here know how to sail. He didn't,' agreed fellow local ex-kid Louise Thompson. 'He'd tip over … I can remember some of the local boys going out to rescue him.'

Einstein was in fact so bad at his favourite hobby that he was literally known across the Eastern Seaboard for his nautical mishaps. Four years earlier, he had made the national news after misjudging the tide on his way home and somehow stranding himself in Connecticut, an entirely different US state from where he started out. He was known in Rhode Island as well, having been rescued by locals several times after capsizing or running aground in heavy fog. And each time, like the hero of some strange, relativity-based Greek tragedy, he would get back in that boat and start sailing again, forcing good Samaritans across the coast into the reluctant role of keeping the smartest man in the world from accidentally drowning himself.

Now, the less accommodating reader may well be wondering why people continued to rescue the poor capsized Einstein. After all, there's only so many times you can see a frizzy-haired genius physicist bobbing about helplessly in the ocean before you start to suspect he might be taking advantage of you. But unfortunately for the people of Cutchogue, Einstein had a trump card: this salty seadog had never learned to swim – which, considering how much time he appears to have spent falling into water, was honestly kind of impressive.

Luckily for him, he had built up enough goodwill among the locals to stop them from willingly becoming The Town That Killed Einstein. He was the guy who played music with old Mr Rothman in the department store; the man who would sit on the rocks outside little Martha Paul's backyard and watch the ocean.

'To us, he was just a bad sailor with funny hair and a funny accent,' Paul would later tell *The New York Times*. 'People used to look out there and laugh at this strange guy in his sailboat going nowhere.'

But if the residents of Cutchogue – and Connecticut, and Rhode Island, and everywhere else Einstein was fond of capsizing – had known what his friends knew, they might not have been so forgiving. 'While we were engaged in an interesting conversation I suddenly cried out "Achtung!" for we were almost upon another boat,' his friend and fellow academic Leon Watters once reminisced.

'He veered away with excellent control and when I remarked what a close call we had had, he started to laugh and sailed directly toward one boat after another, much to my horror; but he always veered off in time and then laughed like a naughty boy.'

Yeah. I bet, up until now, there was a part of you that was still inclined to give old Albert the benefit of the doubt. After all, the poor guy may have been an idiot on the water, but it's not like he was doing it on *purpose*, right? But no: whenever he ran *Tinef* aground, or bobbed about in the water watching the locals scramble to his rescue, he wasn't ashamed – he thought it was hilarious.

And although the reluctant lifeguards of Long Island didn't know it, they had the fate of the free world in their hands – or, you know, their lifebuoy rings. For this would be a strange summer for the aging physicist: when he wasn't busy tipping himself into the Atlantic Ocean, Einstein was spending his holiday in direct contact with the President of the United States, discussing the minor matter of the development of a brand-new type of bomb. Europe at this point was just weeks away from all-out war, and yet, strange as it may sound from a modern, post-Team America World Police perspective, the US was intensely uninterested in entering the fray, preferring to leave those ridiculous continentals to kill themselves on the battlefield without their involvement.

But with an influx of refugees from the Nazis came something the isolationist USA couldn't ignore: information. Just the previous year, in 1938, scientists in Germany had become the first humans to witness nuclear fission – that is, how to split the atom.

It was an incredible discovery on its own, but it would extremely quickly become overshadowed. For those people smart enough to understand what had happened and why, the logical next step – especially considering when and where and under which government it had occurred – was to try to figure out how many people it could kill.

Now, scholars of ignorance, which sounds like an oxymoron but bear with me, sometimes classify secrets into one of two types. There are what's called subjective secrets, which are defined as being compact, transparent, changeable and perishable – in other words, short, easily understandable bits of information that will pretty soon no longer be secret in any case. Think, 'the troops will attack at dawn', for example, or 'I'm buying Stinky Jeff some deodorant for his birthday.'

The other type of secret is the opposite in just about every way, including the fact that they're not really secret at all. These are objective secrets: diffuse, technical, determinable and eternal, or, to put it another way, things which last for ever and could, given the right line of enquiry by the right person with the right training, be figured out from first principles. Like how to make an A-bomb.

That's why, normally, if you're the kind of person who knows exactly how to build a nuclear weapon, you'll probably also have been made to sign up to some Official Secrets Act to stop you advertising the fact. It's bad enough that our enemies could theoretically figure out how to do it themselves without our help, people say; we don't need our own scientists going over and giving them shortcuts.

But the good thing about Nazis is that they're fucking idiots. Fully half of the four-person team who discovered nuclear fission were Jewish, and another one was so vehemently anti-fascist that he sheltered a Jewish woman for months during the Nazi regime and reportedly said that 'if my work would lead to Hitler having an atomic bomb, I would kill myself,' which, firstly: badass, and secondly meant that the Nazis had a problem. They had, essentially, discovered nuclear fission, and then immediately driven just about everyone in the country likely to understand how to make it into a weapon into the hands of their enemies.

One of those who had fled Europe for the US was a Jewish Hungarian-born physicist named Leo Szilard. He, too, was in Cutchogue that July – not to sail, but to fight Nazis. He was on a mission to alert the US government to the potential risk of weapons based on nuclear fission, and to convince them to start some kind of nuclear research facility themselves – to get out ahead of Hitler in the race to create an atomic bomb. And, for that, he figured, he needed Einstein.

See, by this point, Einstein was so famous in the US it was practically a superpower. He reportedly suffered being stopped by fans in the street so often that he had to adopt a cunning disguise – 'He says with great humility, and in broken English, "Pardon me, sorry! Always I am mistaken for Professor Einstein,"' recorded the *New*

Yorker in 1939. 'People turn away without saying any more' – and later on in his life, when the US Government attempted to indict the sociologist, historian and civil rights activist W. E. B. Du Bois under trumped up charges that basically amounted to 'he's Black and a leftist', Einstein's offer to appear as character witness got the case thrown out mid-trial just through sheer force of reputation.

That popularity hadn't always been the case for Einstein: when his theory of general relativity was first confirmed in 1919, the initial reaction in Europe was to celebrate – it was, according to J. J. Thompson, President of the Royal Society, 'one of the momentous, if not the most momentous pronouncements in the history of human thought'. But in the USA, a country struggling with paranoia over communism, staggering wealth inequality and rampant xenophobia, things were very different. Relativity was presented – in *The New York Times*, no less – as 'Bolshevism' and the kind of thing that would make 'the Declaration of Independence itself ... outraged'.

And then suddenly, that all just kind of ... flipped. When Einstein arrived in the US for the first time in 1921, it wasn't as a physicist, but as one of a group of delegates of the International Zionist Organisation. That detail is important, because it's what led to the wacky series of cultural mix-ups that made Einstein such a celebrity: New York, where the ship carrying him and his fellow delegates docked, was home to around 1.6 million Jews at the time – about half a million more than today – and they turned out in their droves to see the arrival of one passenger in particular: Chaim Weizmann, the President of the IZO.

The mainstream press, however, who mostly hadn't heard of Zionism or Weizmann, just saw huge crowds gathering to see a boat carrying a bunch of guys from Europe, and one of them was Einstein, the guy who had just recently turned science communist or something. They didn't know who the other fellas were, so it stood to reason that everyone was there to see Einstein. Wow! Who knew New Yorkers loved physics so much?

From then on, Einstein's image in the popular American consciousness only grew. The public learned that, far from being the haughty European brainiac people had assumed, the professor was actually a kindly old grandpa-like figure, who smoked a pipe and projected an air of polite bafflement at his own fame.

And at the same time, Einstein was being denounced by the communist government in Russia as being 'reactionary' and 'supporting counter-revolutionary ideas'. He was forced to flee Germany temporarily after receiving death threats from antisemitic hate groups, and after Hitler seized power in 1933 he left for good – the Nazis put a $5,000 bounty on his head and published his image under the words 'Not Yet Hanged', which is the kind of thing that can really make a person feel unwelcome.

So Einstein came to America as a refugee, leaving a country where Jews were banned from holding professorships anywhere for a country where Jews were banned from holding professorships *almost* anywhere.

Yes: this was back in the days when racism wasn't so much dogwhistled as it was shouted proudly from the rooftops – if you've ever wondered why such gigantic intellects as, say, Jonas Salk or Richard Feynman never attended any of the Ivy League colleges, the reason is as simple as it is depressing: they were Jewish. The Ivy League didn't want Jewish students – and neither, for the most part, did they want Black students, or women, or really anybody stupid enough to have been born with melanin instead of a trust fund. One of the very few places in the US without racial quotas at the time was the Institute for Advanced Study, an independent centre for research, still less than a decade old, in Princeton, New Jersey. At a time when admissions panels at Yale were still being advised to 'never admit more than five Jews, take only two Italian Catholics, and take no Blacks at all', the IAS had been founded with zero restrictions on race or sex or religion or nationality; it was, in its founder's words, 'an educational Utopia'.

For Einstein – one of the few mainstream figures before the Civil Rights era to point out the racism in American society – it was perfect. 'He was kind to all young people, but totally uninterested in Princeton society,' writer Jean Shriver reminisced about her famous ex-neighbour in an article many years later for the *Daily Breeze*, an LA County daily newspaper. 'He ignored the unwritten racial segregation of the town and became one of the few white people to wander through the black section, where he handed out nickels to children.'

And he used his fame for his cause as well. In an address at Lincoln University, the first college in the US to grant degrees to Black students, he called racism 'a disease of white people', which he '[did] not intend to be quiet about'. He worked with his friend Paul Robeson on the American Crusade to End Lynching, he was a member of the NAACP, and he was known about Princeton for quietly heroic acts like offering his home to Black travellers when they were refused rooms at local hotels and personally paying for a Black kid's college education – all dangerous behaviour that could get him on a watchlist, and in fact did: by the time Einstein died, the FBI had collected nearly 1,500 pages of intel on him for his social justice work, which is approximately 1,500 pages more than they had on, say, the Ku Klux Klan.

Which is why the letter written by Einstein and Szilard warning of the Nazis' potential to build a nuclear bomb put President Roosevelt in such a bind. The information was good – and even if it wasn't, if it got out that he had ignored Professor Einstein's advice on something of this enormity, the public would never forgive him.

But, equally, he didn't trust this little German commie. Sure, he seemed to love his adopted country, and, yes, the *actual* communists had formally denounced him, but all that organising for racial justice? That just wouldn't do. So he decided on a compromise: while the Einstein-Szilard letter has gone down in history as the blue touchpaper that started the Manhattan Project, ultimately ending in the nuclear destruction of Hiroshima and Nagasaki, Einstein himself –

probably the one person on Earth you'd most expect to be involved – was officially banned from working on the A-bomb.

It may have been for the wrong reasons, but with that decision the US government may have dodged a major bullet. Remember how Einstein had that habit of, you know, barrelling head-first into situations with no thought as to any potential dangers or consequences? Well, nowhere was this more true than in his love life.

Now, given his refusal to accept social mores, Einstein was not obvious boyfriend material. He dressed like a weirdo, eschewing socks; he hated baths; and he was known for waking his landlady and neighbours up at obscene hours shouting 'It's Einstein; I've forgotten my key again.' In short, you probably wouldn't want Einstein as a personal friend. And you *certainly* wouldn't want him as a husband.

Just ask Einstein's first wife, Mileva Marić. Like so many couples, the pair met at university, in the maths and physics department of what is now ETH Zurich, and in a just world we would all know the name Marić with the same familiarity as we do Einstein. She was, by all accounts, the better scientist: she scored higher grades than him – significantly so when it came to applied physics, where she got the top score and he the lowest – she was the class swot and he barely turned up to lectures, and she went on to get a professorship while Albert did not. And without Marić, there's a pretty good chance we wouldn't have got relativity at all, since there's a whole heap of evidence that she was the unlisted second author of Einstein's work on $E = mc^2$. In Einstein's words, 'I need my wife. She solves for me all my mathematical problems.'

And what did she get in return? Her first child given up to adoption, her discoveries renamed as her husband's and a marriage full of infidelities.

'A. You will see to it (1) that my clothes and linen are kept in order, (2) that I am served three regular meals a day in my room. B. You will renounce all personal relations with me, except when these are required to keep up social appearances.'

This was the letter Mileva received in 1914 from her husband and father to her three children – one of whom, a twelve-year-old daughter named Lieserl, he had never seen: having dragged his feet into marriage with Mileva, Albert had forced his girlfriend to give the baby up for adoption to save face. The pair had eventually married in 1903; Einstein's *annus mirabilis* – the year he published his three most famous and fundamental papers, one on photoelectric effect, another on Brownian motion, and the final on relativity – came in 1905, and by 1912, with Mileva busy looking after their two young sons, Albert had started playing away with another woman. As if following some unwritten law of genius cads throughout the ages – and no doubt inspiring comedians the world over to shout 'not that kind of relativity, stupid!' – that woman was his cousin.

'I treat my wife as an employee whom I cannot fire,' Albert wrote to Elsa. 'I have my own bedroom and avoid being alone with her.'

And to Mileva, he commanded that 'You will expect no affection from me ... You must leave my bedroom or study at once without protesting when I ask you to.'

Unsurprisingly, the couple split – although it took four years of estrangement and a nervous breakdown on Mileva's part before the divorce went through.

Albert married Elsa, and for a long time everybody forgot about Mileva completely. It was Elsa who accompanied him to the US, so it was Elsa who got that reflected fame. But as mothers across the world tell their children: if a man will cheat with you, he'll cheat on you, and Albert was by no means a better husband to Elsa than to his first wife.

So there was Estella. There was Ethel. There was Toni, and the enigmatic 'M' and 'L'. There was the woman who, he wrote to Elsa, 'acted according to the best Christian-Jewish ethics: 1) one should do what one enjoys and what won't harm anyone else; and 2) one should refrain from doing things one does not take delight in and which annoy another person. Because of 1) she came with me, and because of 2) she didn't tell you a word.' All of these women he would describe in letters to his cousin-wife, telling her about all the fun they were having together over in Europe while she was stuck in the US. He even roped in Elsa's daughters to help him pass love letters to them.

And then there was Margarita Konenkova. Einstein first met the Russian émigré in 1935, when her husband Sergey was commissioned to sculpt a bust of the famous physicist. It's not known when their acquaintance turned into a full-blown affair, but they were definitely an item by the mid-1940s, as a collection of love letters that turned up in the late 1990s confirmed.

'Everything here reminds me of you,' Einstein wrote to Konenkova in a letter dated November 1945. 'Almar's shawl' – 'Almar' being a portmanteau of 'Albert' and 'Margarita', proving that the couple really were the Bennifer of their day – 'the dictionaries, the wonderful pipe that we thought was gone, and really all the many little things in my hermit's cell; and also the lonely nest.'

'Be greeted and kissed, if this letter reaches you, and the devil take anyone who intercepts it,' he signed off in another, dated three months later in February 1946. 'Your A. E.'

Here's the problem, though: the world was just about to enter the Cold War, and Margarita was suspected of being a Soviet spy.

To this day, nobody really knows for sure whether the rumours were true – spying does, by its very nature, tend to be somewhat clandestine – but her family certainly seemed to think it was true when they told *The New York Times* in 1998 that 'she was the number one spy for the Manhattan Project'. Pavel Sudoplatov, a former Soviet spymaster who wrote a tell-some book in 1995 about his work in intelligence, also pointed the finger at Margarita, listing her mission as 'to influence [Manhattan Project laboratory head Robert] Oppenheimer and other prominent American scientists whom she frequently met in Princeton', and to introduce Einstein to Soviet vice consul Pavel Mikhailov.

Spy or not, she succeeded in at least one of those objectives – Einstein references meeting Mikhailov in some of the surviving letters between the two. And yet, despite her habit of using sexy talk like 'come with me to the Soviet embassy and meet my comrade Pavel, babe, he's really interested in nuclear weapons', Einstein apparently never suspected that his girlfriend might be a foreign agent.

Which just goes to show, really: you may have the biggest brain on the planet, but it doesn't matter if you're thinking with a different organ entirely.

26.

Kurt Gödel, the Disney Princess Who Broke Time

You know what? I'll say it: the dark ages get a bad rap. Sure, most people couldn't read, and there was a better than even chance your life of perpetual serfdom would be cut short when you died from the plague aged nineteen, leaving your wife and nine children to make a life for themselves in the woods with ye Merrie Bande of Outlawes. But at least we knew our place in the universe: right in the goddamn centre. Humanity was, very literally, what everything revolved around; the very point of creation itself, and while we may not have understood the finer points of things like physics or maths or basic hygiene, we could rest safe in the knowledge that *Somebody* did, because the village priest had told us so.

Then the scientific revolution came along and fucked everything up. Suddenly, everything was difficult and confusing: we had been demoted to some random planet revolving round some random star in a potentially infinite universe, and if we wanted to understand even a small part of it we would have to start doing things like 'performing experiments' or 'inventing calculus'.

Kurt Gödel, the Disney Princess Who Broke Time

Fast forward through three or four hundred years of painstaking eye-poking and turkey electrocution, and humanity had finally got itself back on its feet. By the end of the nineteenth century, our corseted forebears had figured out, among other things: fluid dynamics, psychoanalysis, the theory of evolution, telephones both for the living and the dead, *and* opera rollerblades, and they were justifiably pretty smug about it. There were just a few odds and ends that needed tidying up – David Hilbert's famous list of twenty-three unsolved maths problems, for example, or what the heck X-rays actually were – and science would officially be Finished.

And then, like a little fairy tale imp, up popped the twentieth-century Austrian mathematician Kurt Gödel to ruin everybody's day all over again.

In school, when you asked your weird nerd friend why they liked maths more than, say, English, there's a fair chance they'd hit you with something along the lines of 'I like the security of it.' You know what I mean: there's always an answer, it's either right or wrong, no waffling about whether Iago secretly fancies Othello or if he's just a massive racist. Now, some of those friends may have gone on to do a maths degree – some may even have become maths professors. Ask them again whether their pet subject has all the answers. I'll wait.

. . .

Back? Right: that exasperated sigh you almost certainly heard before forty minutes of muttering about something called 'incompleteness' is entirely the fault of Kurt Gödel.

Despite what your primary-school teachers told you, there aren't many figures from history who single-handedly changed the way humanity understood the universe. But Gödel actually *did* do that, and the response from the scientific community around him was first to call him a nutcase, then to denounce him as an iconoclast, and then, finally, many years later and on a different continent, to call him a genius.

But what exactly was Gödel's big idea? Kind of a cop-out, on the face of it: basically, he proved that some things can't be proved. Big deal, you might think, there's a *bunch* of things that I can't prove – do *I* get to be a genius? But that's not what Gödel meant: he proved that there are some mathematical hypotheses, including a few that underpin the very fabric of maths itself, which can't *ever* be proved. By anybody. Ever.

So, as you can imagine, this revelation kind of … broke mathematicians. Centuries – nay, millennia – of patient study and obscure intracontinental notation wars just trying to get back to the warm comfort of knowing that everything could be figured out, and along comes some twenty-four-year-old from Vienna with an ex-nightclub dancer divorcée girlfriend to tell you some of the most important ideas ever posited by a human brain will just *never be answered*. Even the great Bertrand Russell – he of floating space teapot between Earth and Mars fame – couldn't wrap his head around it, saying the theorem would mean 'that 2 + 2 is not 4, but 4.001', and that he was 'glad [he] was no longer working at mathematical logic'.

Now, it should be said that Gödel himself took completely the opposite view. As far as he was concerned, he hadn't broken maths, he had simply shown that it transcended reality. By proving that there were some ideas that could never be solved using logic, he had shown that humanity was special – that we could never be replaced by computers or algorithms, because they lack the kind of intuition that allows us to parse things like 'this sentence is false' without having our heads explode.

But listen: we're not here to talk about how Gödel revolutionised maths – there are textbooks for that. So, instead, let's talk about Kurt Gödel, the paranoid flamingo enthusiast whose favourite film was *Snow White and the Seven Dwarves*.

See, the thing about Gödel is: you really get the sense he *wanted* to enjoy life. He was an inquisitive kid – his parents called him Herr Warum, or 'Mr Why', because of the number of times he asked the question. He was clearly a romantic, too: he fell in madly love as a young man with a cabaret dancer named Adele; she was six years his senior and had been married once already, and his strictly religious parents *hated* her. His theorems and hypotheticals may have frustrated everyone around him, but he only ever presented them with a kind of awed reverence for the Truth they revealed. And he was silly and frivolous, in ways that seem entirely wrong for his neat image and highly abstract reputation – students who took the time to look up his address and sneak over to see the Gödel residence in Princeton would be taken aback when they came across a big, bright pink flamingo lawn ornament, which Adele had planted outside Gödel's study and he had adored ever since.

But at every point in his life, demons loomed. He had always been a nervous kid, obsessed with illness and disease ever since a bout of rheumatic fever in early childhood – he had completely recovered from the problem in every way except mentally, and was left with a lifelong conviction, based on no evidence other than his own anxieties, that his heart had been permanently damaged. As an adult, he would take medication daily for this phantom cardiac condition, and in fact he took meds for all kinds of illnesses: according to one of his closest friends, Oskar Morgenstern, by the end of his life Gödel was taking a daily cocktail of treatments for digestive, bowel and cardiac problems he didn't have, as well as for kidney and bladder infections he didn't have, which goes to show not only how disabling his health anxiety was, but also just how lax pharmacists' licencing laws were back in the '70s.

But the paranoia really began acutely in June of 1936, when Gödel's hero and mentor, Moritz Schlick, was murdered on his way to a seminar. It was Schlick who had first piqued Gödel's interest in logic, and his death – purportedly at the hands of an ex-student driven into a murderous rage by too much philosophy – threw Gödel into 'a severe nervous crisis', according to his brother, for which he had to be committed to a sanatorium for a while.

Two years later, after the Nazis annexed Austria, Schlick's killer would receive a full pardon for the crime. He had, he argued, 'by his act and the resultant elimination of a Jewish teacher … rendered National Socialism a service'. Schlick was not in fact Jewish, but that didn't seem to matter; he had studied and taught on philosophy and relativity, both suspiciously Jew-y topics in the eyes of Nazis, and that was enough of an excuse for his getting literally shot in the street to be just fine with the authorities.

Gödel, also not Jewish, fared better than Schlick under the new regime, but his dual interests in logic and showing the sum of all received wisdom to be built on flimsy assumptions had made him a marked man – not to mention his distaste towards violent antisemitism. And so, despite having already proven results that previously had never even been conceptualised, let alone attempted, Gödel found himself unhireable as a mathematician in Austria. Not only that, but the newly emboldened fascists had developed a nasty habit of beating him up in the street, with his only defender being Adele and her umbrella.

That wasn't to say he had *no* prospects in Vienna, and in September of 1939, he received a fabulous, once-in-a-lifetime offer of conscription into the Wehrmacht to go fight in the Second World War for the same guys who had murdered his mentor and physically attacked him in public. His reaction was to do precisely what we'd all do in such a situation, which is to elope with Adele and then quit Germany so hard that they got all the way through the entire continent of Asia and ended up on the West Coast of the USA. From there, the pair then boarded a train to Princeton, where Gödel accepted a job at the Institute for Advanced Study alongside a fellow refugee from the Nazis – an aging German-born physicist named Albert Einstein.

Einstein and Gödel were different in almost every way possible, but they got along like a house on fire – in fact, Einstein, by this point already one of the most famous people in the world, once said that the only reason he still went to his office each day was 'to have the privilege of walking home with Kurt Gödel'.

And let's face it: if anybody would be immune to the kind of 'but what if everything you thought ... wasn't?' logic Gödel specialised in, it would be Einstein, right? Dude could out-think anybody, and frequently did. But Gödel's meddlesome brain spared nobody: not Gödel himself, and certainly not Einstein.

It had been more than three decades since Einstein had given the world the theories of relativity – longer than Gödel himself had been alive, in fact – and by this point, people were starting to work out the consequences of this radical new worldview. Consequences like: the Big Bang Theory, which, aside from being an inexplicably successful sitcom, is the current leading explanation for the existence of the universe. You've probably heard of it; it's kind of a big deal.

What you may not have been taught in school is that a 'big bang' followed by rapid expansion isn't the only version of the universe that makes sense theoretically. It wasn't even the version Einstein himself preferred. He favoured the 'steady-state' model, in which the universe is eternal, with constant density, and matter in constant creation.

Before we discovered practical evidence for the big bang theory – like cosmic background radiation, which is rather difficult to explain without a single origin point for the universe – you could pretty much believe either of these theories without attracting too much scientific derision. But Gödel went a third way: he created a cosmological model that made perfect sense according to all established laws of physics and mathematics, and the only problem was that it made absolutely no goddamn sense in any other way.

The issue with Gödel's model – at least as far as Einstein saw it – was that he had accidentally discovered time travel. Or ... well, maybe he proved that time doesn't exist? Or that it was actually just the same thing as space ... it was all very 'disturbing', Einstein thought. Basically, in Gödel's universe, it would be possible to travel to any point in time just by going far enough through physical space, and unlike his contemporaries who simply thought it was a cool little observation that time travel isn't ruled out by the laws of physics,

Gödel saw in his bizarre thought experiment a deep revelation about the nature of the universe. If a point in time can be revisited, he reasoned, then it can't really be said to have 'passed' in any meaningful way – the only conclusion to be drawn was that time didn't exist at all. And if it didn't exist in Gödel's weird hypothetical universe, then it couldn't exist in any, he thought – including our own.

Having now successfully broken the laws of logic and time, Gödel set out to ruin something even more important: the US constitution.

Now, I'm no lawyer, but I have seen a whole bunch of *90 Day Fiancé*, and I've learned that if there's one point in your life that you want to actively avoid pointing out the shortcomings of the US constitution, it would probably be 'right in the middle of your citizenship hearing'. Gödel, honest and logical as he was, did not get this memo.

The problem was, he cared too much. He knew he would have to pass a citizenship exam, as his sponsors Einstein and Morgenstern had done previously, and so this 'very thorough man' spent months before the hearing 'informing himself about the history of North America by human beings', Morgenstern would later recall.

But, then, something terrible happened. He'd done it again. 'He rather excitedly told me that in looking at the Constitution, to his distress, he had found some inner contradictions,' Morgenstern said. Gödel had discovered that he 'could show how in a perfectly legal manner [how] it would be possible for somebody to become a dictator and set up a Fascist regime, never intended by those who drew up the Constitution.'

His friends, naturally, told him to forget about it – just chill out, tell the judge you're hot for Uncle Sam, and everything will be fine. And maybe it would have been, if only the judge hadn't been in the mood for a chat.

In the end, Einstein had to forcibly tell Gödel to shut up after he had started to tell the judge that the constitution wasn't worth the paper it was written on. But even more confusing than how anybody is admitted into the US at all, let alone as a full citizen, immediately after pointing this out, is what those 'inner contradictions' actually are. To this day, nobody has been able to work out what bit of the constitution Gödel had in mind when he rang his friend in a panic that night, and 'Gödel's Loophole', as it's now known, has continued to haunt the nation's foundational document ever since.

In 1955, Einstein died, and Gödel's main link to the outside world was gone. Other than Adele, Einstein was the only person Gödel spoke to regularly, and without this warm and gregarious companion to walk him home each day, he became even more withdrawn and paranoid. He started seeing conspiracies in TV listings, which is generally a bad sign as far as mental health is concerned, and he became convinced his co-workers were plotting to kill him – like his beloved heroine Snow White, he thought, he was to be poisoned, and so he stopped eating just about anything except butter, baby food and laxatives, which actually sounds even worse than poisoned apples. He refused almost all conversations that weren't by telephone, even with people just in a different room, and if somebody wanted to talk to him in person, he would propose a very specific time and place to meet and then make sure he was as far away as possible. And, normally, this would be a sensible policy for a happy life, but Gödel's introversion was so extreme that he stood up the President in 1975, which when combined with his shenanigans in his citizenship ceremony makes you wonder why there wasn't an FBI agent permanently stationed in his back yard.

But perhaps the reason he got away with these kinds of missteps was that he was just so … *innocent*. He wasn't pushing boundaries – just as when he had upended maths and physics, he seemed blissfully unaware – or at least, fraughtly unaware – that his actions could piss anybody off. He was the kind of guy who would happily tell Einstein that time travel was possible, and would assume Einstein would be happy to learn the mind-boggling implications of his theorem – and because of that, he was the man Einstein made time for every day.

In 1977, his beloved wife Adele was hospitalised, and Gödel stopped eating altogether. She was the only person in the world he trusted not to poison him, and without her to make and test his food, he simply starved. He was admitted to Princeton Hospital in late 1977, weighing just four and a half stone and looking, according to friends, like 'a living corpse'. He died there two weeks later, in January 1978, with his death certificate recording the cause of death as 'malnutrition and inanition' resulting from a 'personality disturbance'.

Poor Kurt Gödel. Honestly, if it wasn't so tragic, it would have been quite ironic that the man who transformed the world of logic died in such a paradoxical way – starving to death out of fear of being poisoned – but, instead, it's just a terribly sad way for one of the most incredible figures in maths and science and philosophy and flamingo appreciation to die.

But perhaps we shouldn't mourn unduly. In Gödel's universe, after all, time is an illusion, and mathematical truth transcends the physical world. Maybe, if we just travel far enough in the right direction, he'll still be out there doing it all again – and maybe, this time, we can take along a whole bunch of antipsychotic meds as well. Maybe see if he can fuck up Stephen Hawking's day too.

27.

Maya Angelou, in: Stop! Or My Mom Will Shoot

Maya Angelou once said that she had 'never been bored in [her] life', and that makes sense, because she also once said 'who needs three days of rest? Please! The second day, you might die.'

In other words, the woman got a lot done in her eighty-six years on Earth. To be honest, if you tried to tot up in a notebook all the experiences she managed to cram into her life, you'd run out of paper before she reached forty.

Seriously. If you don't know all that much about Dr Angelou, you probably think of her as a writer or a poet, or at the very least an occasional side-character in *The Simpsons* – and she absolutely was all of those things, penning no fewer than seven autobiographies, eighteen volumes of poetry, three memoirs, seven plays, dozens of TV shows and films, a handful of children's books and – just showing off now – two cookbooks. She was a journalist in Africa, and she wrote the first ever screenplay by a Black woman to make it to the big screen; she wrote one poem that earned her a Pulitzer nomination, and another which was read aloud at the inauguration of President Bill Clinton in 1992 … by her.

But to just call Angelou a 'writer' is to do her a disservice. She was also a leading civil rights activist – as one obituary put it, not just a participant in the struggles of the 1950s and '60s, but on staff. Throughout her life she knew or worked with pretty much every big name you've ever heard of in relation to the fight for Black liberation: Martin Luther King Jr and Malcom X counted her among their friends and co-organisers, and she had met Nelson Mandela while living in Cairo – her books would later keep him company throughout his imprisonment on Robben Island. She marched for women's rights with Gloria Steinem, and was an early advocate for LGBTQ+ rights.

And that's not all. She was also, at various times in her life: a Tony-award nominated actress; a calypso star; a triple-Grammy award-winning recording artist; a Presidential Medal of Freedom recipient; a professor of American Studies at Wake Forest University in North Carolina, with at least thirty honorary degrees under her belt; a sex worker; a pimp; and a mother, not just to her own son, but also to the 'daughters' she had across the world: 'sometimes I'll get a thousand pieces of mail a week from young women who think I'm wise', she once said. 'They use me as a mother and I think of them as my daughters.' And just to rub it in a bit further, she did all of this in *six* languages. The woman couldn't even fail GCSE French like the rest of us plebs.

Given all this activity, it does mean that her accomplishments tend to overshadow some of the stories from earlier in her life. But we really should hear about them, because what Cliffs Notes biographies like the above can't convey is just how bonkers the details of her life were.

Like: we tend not to hear about Maya Angelou the teenage San Francisco streetcar conductor. It makes sense – it was only a few months of her life, and it doesn't exactly sound like the kind of adventure that could rate with hanging out with Nelson Mandela in terms of world-changing-ness – but it's definitely a shame. After all, she may have been unmatched as an adult, but as a kid she was incredibly relatable: she liked cute outfits, she was willing to annoy her way into getting one, and she was – as, trust me, all of us should be – terrified of the wrath of her mum.

Now, Maya Angelou didn't really know her mother at all as a young kid. And I don't mean that in a sort of 'who really *knows* their parents, man?' kind of way – she *literally* didn't know her mother. Aged three, she and her older brother, Bailey Johnson Jr, had been sent more than 1,500 miles east to live with their grandmother in rural Arkansas, and there are at least two things in that sentence that might strike you as odd. Firstly, you'll notice her brother had the surname Johnson rather than Angelou – but, in fact, at this point so did she; Maya Angelou's birth name was Marguerite Johnson, and for a long time she was known as 'Rita' or 'Ritie' in public. 'Maya' was actually a nickname that came from her brother's baby-talk pronunciation of 'My-a sister', which stuck so hard she eventually chose it as her pen name.

The other thing to note is that, yes, the siblings were *sent*, not taken, to their grandmother's house in rural Stamps, Arkansas: according to her first autobiography, *I Know Why the Caged Bird Sings*, the pair had been 'shipped home' aged three and four, relying on just the train porter and the (hopefully) good will of their fellow passengers to get them safely across the country. They were packaged on to the train like evacuees, or, you know, luggage, with their tickets pinned to Bailey's coat and tags on their wrists listing their names and destination, and, incredibly, despite the train porter getting off the train the very next day in Arizona, the journey was a success.

Stamps would be where most of the young Marguerite's earliest memories were formed. It was a town she would later recall as being 'with its dust and hate and narrowness … as South as it was possible to get', which in the 1930s meant that her disabled uncle would sometimes have to hide in the potato bins to avoid being lynched. 'People in Stamps used to say that the whites in our town were so prejudiced that a Negro couldn't buy vanilla ice cream,' she wrote, and recalled it being so starkly segregated that she didn't fully believe white people were real.

It wouldn't be until she was about seven that she would be reunited with her parents again, when her dad suddenly turned up out of the blue and drove the siblings nearly 500 miles north to Saint Louis. That was where Angelou's mum lived, and it was love at first sight for little Ritie: 'I was struck dumb,' she wrote. 'I knew immediately why she had sent me away. She was too beautiful to have children.'

Now, you might think it would be difficult to understand what life must have been like for the young Maya Angelou, newly transplanted into her mother's family in an unknown city. But, really, if you've ever seen a British soap opera or an episode of *The Sopranos*, you kind of get it. There was her grandma, the German matriarch with white skin and an informal command over the local cops; her mum, the beautiful and elusive Vivian Baxter, with her red lipstick and her mean boyfriend Mr Freeman; her uncles, all four of them, known around the city for their fierce loyalty to their family – and their reputation for violence against anyone who crossed it.

Baxter may have been a head-turner, but mother of the year she was not – at least when Angelou and her brother were small. Now, in fairness, it's not everybody who's cut out for looking after young kids – toddlers in particular rank very high on the scale of 'things you wouldn't think would be that hard to do, but Jesus CHRIST' – but Angelou would later say, in her trademark blunt style, that Vivian was actually pretty terrible at the whole maternal instinct gig, and that putting her infant daughter on a cross-country train was possibly the best parenting decision she could have made.

This first reunion between mother and daughter didn't last long, because shortly afterwards Maya was violently raped and sexually abused at the hands of Mr Freeman, and ended up first in hospital, then court. Freeman was found guilty of raping the eight-year-old Maya, but sentenced to just one day in prison, and that, in the Baxter brothers' eyes, was no justice at all. It wasn't long before Freeman was found dead, seemingly by an act of God: he had suddenly and tragically contracted a terminal case of being beaten to death down at the local slaughterhouse.

Weirdly, when Grandma Baxter was told about this terrible misfortune, she didn't seem very surprised, and instead immediately started handing out lemonade and banning the mention of Freeman's name under her roof from now on, which is, I think you'll agree, not suspicious at all.

Maya and Bailey were scared witless, but for different reasons. Bailey, like you'd expect from any nine-year-old, was taken aback by both the talk of death and his grandmother's reaction to it – but, for Maya, it was even worse. She had spoken up about the rape, and now a man was dead. With only a child's grasp of logic, she put two and two together as best she could and drew an obvious conclusion: that her words brought death, and she should stop speaking from now on.

So she did.

At first people were understanding. It was a trauma response from a young kid, and her family were there for her in all the normal and non-incriminating ways you'd expect. But eventually, Maya and Bailey were sent once again to Arkansas, with no reassurance as to whether they were wanted in Stamps, or simply no longer wanted in Saint Louis.

Either way, the pair lived with their grandmother in Stamps for the next five years or so, until what was probably an equally traumatic event prompted them to be sent away again – Bailey, still only a child, had not only seen a lynching victim being dredged, bloated and rotten, from the lake, but had been made to carry the corpse through town to the prison cells, where the local cops had locked him in with the body as a 'joke'.

The kids went in the other direction this time, though – first Los Angeles, then San Francisco, California, where their mother had moved to live with her new husband. And it was from here on out that Maya and Vivian became much closer, and our streetcar story begins.

Now, ever since California had joined the Union back in 1850, it had been seen as something of a promised land for Black Americans – especially those living in the South. It wasn't just because of the gold in them thar hills, although that was definitely a draw – it was the fact that a Black person had *exactly as much chance* of finding that gold as a white person.

'This is the best place for black folks on the globe,' wrote one Black prospector in a letter to his wife back east during the height of the gold rush. 'All a man has to do is to work, and he will make money.'

In California, Black people had a chance to get in on the ground floor of society – not as enslaved people but as community leaders and organisers. And while yes *of course* they still faced racism – this was nineteenth-century America, for goodness' sake; one of the Supreme Court Justices of the new state was actively trying to make California a white-only state – they also made some important strides towards equality.

Streetcars were one of those strides. A couple of court cases in 1866 had secured a ban on segregation in the vehicles, which for perspective is nearly a full century before Rosa Parks was arrested for refusing to give a white man her seat on that bus in Alabama. So it was illegal, when Maya Angelou moved to San Francisco in the early 1940s, for the city's Municipal Railway service to deny her a seat on the iconic trolleys.

But the young Ms Johnson wanted more than that. She wanted a job as a conductor, and it was for one extremely sixteen-year-old girl reason: the uniforms.

'I'd seen women on the streetcars with their little change belts, and their caps with bills and their formfitting jackets,' she would later recall in an interview with Oprah. 'I loved the uniform! So I said, "That's a job I want."'

But there was a problem. Black people couldn't be kicked off the streetcars, sure, but they apparently couldn't be hired on them either – there had never before been a Black conductor in the city.* And Maya

---

* Well, to be completely accurate: there had never before been a *known* Black conductor. When the newspapers reported that Angelou had been hired – because, yes, it was *that* big a piece of news – another conductor took himself

was a teenage girl, which is, as any teenage girl will tell you, probably the hardest and most misunderstood thing in the *world* to be.

It would have been bad enough if they had turned her application down – you know, you didn't get it, but at least you took a risk and went after that cute little button-down blazer, and that's not nothing. But they didn't even let her get as far as being rejected. When she asked to apply, they told her she couldn't, because she was Black, and get out of our office. Which is actually impressive, because it's hard to pull something so dickish that it actually makes you grateful for those jobs that just eat your carefully crafted CV and cover letter and then never contact you again.

Well, Miss Vivian Baxter wasn't having any of that. She commanded her daughter to march herself on down to the office first thing in the morning – before the employees even started their day – and wait until she was given an application form. To pass the time, she advised, Maya should 'sit there and read one of your thick Russian books' like Tolstoy or Dostoyevsky. You kind of get the feeling it was like a first date, in that you're like fifteen and your mum has sent you out looking 'so smart! oh my little baby, all grown up!' and told you to 'just be yourself!', except that in this case the date is with hundreds of years of institutionalised racism and the potential first kiss at the end of the night is a minimum-wage job that requires you to get up at the crack of dawn and deal with grumpy commuters.

And it wasn't easy. Two weeks Maya Angelou had to spend, day in, day out, sitting on the floor of the San Francisco Municipal Railway offices reading gloomy Russian literature while the people who worked there – including secretaries she *literally knew from school* – pranced around doing and saying the kinds of things that would get you banished to an episode of *Mad Men* if you did them today. But still, she stayed.

---

down to the transit office to say actually *he* had been the first Black conductor, except he had just been passing as white for twenty years. To which the Municipal Railway replied, 'OK. You're fired.'

Why? Because she was scared of going home and telling her mum she hadn't.

Can you *actually imagine* how intense Maya Angelou's mother must have been? I mean, it was hard enough to pressure me into doing my homework when I was sixteen, *let alone* perform a two-week one-girl sit-in in defiance of a century of discriminatory hiring practices while facing down racist epithets and derision from a bunch of adults.

Especially when you realise that, even though her mother would tell people it was her 'dream job', Maya Angelou didn't seem to be *that* into the idea of working as a streetcar conductor. When the manager eventually came out to offer her an interview – and he did, which just goes to show that stubbornness can get you anywhere in life, or at the very least into the hiring office of a local transport authority – she told him the reason she wanted the job was simply that she 'liked the uniform and liked people', not that it was some higher calling given to her from a benevolent tram god.

All right, job done, you might think; mother placated. Nope. Now Maya had got the job, it was time for her to slip into something more officially mandated and get to work – and what a first day it must have been. Remember when you were a kid, and your annoying parents would make a racket outside your bedroom at seven in the morning to get you up for school? Well, no offence, but your parents don't have shit on Vivian Baxter: she got Maya up every morning at *4 a.m.* to drive her daughter across town to get to her streetcar job, which seems rather like driving to the gym in terms of wasted opportunities, and she'd stick around for Maya's whole shift. And maybe you're wondering how she did that, given Maya's job was spent constantly on the move. Well, her mother kept a close eye by literally being close to her, in a car, following the tram through until the shift was over, with a handgun on the passenger seat – which you have to imagine must have made quite a few morning commutes just *incredibly* stressful.

Now I don't know about you, but my mum would have given up at the 'get up before 4 a.m.' part. But Miss Vivian Baxter did this routine every day that Maya Angelou worked that job. And the gun wasn't for show, either – Miss Baxter may have been a nurse by training, but she had far more experience running poker dens and running with The Notorious St Louis Baxter Boys. In short, she was a woman who had zero qualms about doling out righteous vengeance upon anybody who threatened her family, and according to her daughter, she had basically an entire arsenal at her disposal to back that threat up.

'Vivian Baxter had ammunition of every sort,' Angelou remembered. 'Little bitty Italian pistols. And .45s and .38s. German Lugers. Everything.'

This wasn't an idle threat. Suppose, for example, you were some total bastard who was dating Maya Angelou, a young breakfast fry cook and aspiring calypso singer in early 1950s San Francisco, say, and you, being a total bastard, decide to accuse her of infidelity, then beat the living shit out of her and kidnap her for an as-yet-undecided length of time.

Let me tell you why that's a bad idea: Vivian Baxter. When she learned what had happened, she went full Avenging Angel mode, and we're talking 'Angel' in the sense of 'Hell's Angels' or 'Buffy the Vampire Slayer's soulless ex who tried to kill everyone a bunch of times', because instead of doing what *your* mum might, and calling the police, Miss Baxter played to her strengths – which, if you remember, mostly involved stuff like 'being armed to the teeth' and 'knowing how to convince terrible abusive boyfriends to tragically slip and accidentally kick themselves to death'.

So Vivian formed a posse. She owned a pool hall at the time, and she got all the biggest, meanest, toughest-looking guys who were hanging around doing whatever it is people do in pool halls – which, based purely on movies, I'm assuming is entirely smoking and scowling – and she took them straight to where Maya's now ex-boyfriend was keeping her hostage and *kicked the fucking door down* to save her baby.

And by this point, Maya was so bruised and broken by the abuse she'd been subjected to that her mother fainted at the sight of her – she had broken ribs and split lips, she had been beaten into unconsciousness multiple times and hit with planks. But remember: Vivian Baxter was a nurse, and more than that, a total badass, so she not only took her daughter home and helped her recover, but when she was asked whether they would be pressing charges on the guy, she said absolutely not, and instead went full Godfather and put a hit out on him.

And later, when she got wind of where that bastard was, she handed her daughter a gun and instructed her to get even.

Now, a pushy mother is one thing when she's encouraging you to get an entry-level job for the holidays, but murder is another thing entirely. So it's a testament to the kind of personality Miss Vivian must have had that Maya Angelou seriously considered it: she took her mother's pistol and walked over to the hotel where her ex was staying, phoned him from the lobby with a faked accent and a story about having met him a few nights ago and wanting to meet him again.

It worked, which shows you just how good faith the guy's purported reason for beating up his girlfriend was, and he came down to meet her. And she ... froze. She couldn't do it.

Of course, not being able to murder someone in broad daylight is actually not a personality flaw. In fact, most psychologists would probably say it's a good thing, actually, to not be able to kill somebody – even if your mother *is* probably right when she says that nobody would blame you and you wouldn't see a day in prison for it.

But oh, how disappointed her mother was in soft-hearted Maya. She shook her head: 'You didn't get that from me,' she told her daughter. 'I'd have shot him like a dog in the street.'

Now, they say that behind every great man is a great woman, but the lesson from Maya Angelou's life – well, at least, one of the many lessons we can take from her life – is that the same is true for great women. Angelou was never shy about admitting to the mistakes she made over the years, often describing her many memoirs as a sort of prolonged cautionary tale, like *The Hobbit* but real. But throughout it all, she had her mother behind her – and reading about Vivian Baxter, you really get the impression that's all anybody could have needed.

Who else, in the 1940s, would react to learning that their sixteen-year-old daughter, still in school, unwed – in fact totally single – was thirty-seven weeks pregnant, by drawing a bath and calmly, without judgement, working through the situation?

And who else, when you fail to shoot down in the street the man who kidnapped and beat you, would give you a hug and reassure you that even if you don't have the heart of a stone-cold killer, they still love you anyway, and they'll just have to take him out themselves, it's OK.

Honestly, we should all be so lucky to have someone in our life as loyal as Maya Angelou's mother. And equally honestly, we should all be grateful that Maya Angelou *did*, since she didn't release her first book until she was in her forties. If that ex-boyfriend hadn't been dealt with through the full force of Miss Vivian Baxter, Avenging Angel, the world may well be out a whole oeuvre of culture-defining literature.

'She was a knockout, Vivian Baxter,' Dr Angelou told Oprah. 'Whenever the world would throw me flat on my face ... I would go home to Vivian Baxter.'

28.

Ernest Hemingway May Have Been the Worst Double Agent Ever

In the early hours of 2 July 1961, Ernest Hemingway climbed the stairs from the basement to the front foyer of his house and, with his favourite Abercrombie & Fitch double-barrelled shotgun, took his own life.

For years, he had been plagued with increasing depression and paranoia – illnesses his friends and loved ones put down to his hard-living, hard-drinking history. He was convinced he was being tracked at all times by the FBI, telling his friend that life was 'the worst hell. The goddamnedest hell' because of it.

The last few months of his life had seen the Nobel and Pulitzer Prize winner checked into multiple hospitals and given a dozen or more rounds of electroconvulsive shock therapy to cure his delusions of being on a government watch list. Nothing worked; his last bout of treatment was completed just two days before his suicide.

But here's the thing: *he was right.*

So how did such a quintessentially American writer as Ernest Hemingway end up the subject of a two-decade-long, 122-page FBI investigation?

Ernest Hemingway May Have Been the Worst Double Agent Ever

Look. Ernest Hemingway is another one of those figures from history where a good 50 per cent of what you hear about him sounds completely made up. His life was, undoubtedly, a fucking mess – but in truly the most amazing ways: he was an inept double agent, a guerrilla antifascist, a recalcitrant drunk, a pro-worker and pro-refugee organiser, a legendary war reporter, a four-time husband and cat obsessive and quite possibly some kind of pirate king. Oh – and he might have single-handedly broken the Geneva Conventions at one point.

So when I tell you that his work for the NKVD, the less-memorably-named precursor of the KGB, *didn't even feature* in his FBI dossier, hopefully you can understand the kind of personality we're dealing with here.

It all started in New York, in the winter of 1940. Hemingway was forty years old,* and newly married to his third wife, the legendary war correspondent Martha Gellhorn. They had met in Spain while covering – and, for Hemingway at least, surreptitiously taking part in – the civil war there for the North American press; Gellhorn would soon be moving to China for a new assignment, and Hemingway was going to go with her. And that was when he met Jacob Golos.

Golos had been born in what is now Dnipro, Ukraine, but was then Ekaterinoslav, in Tsarist Russia. He was a revolutionary from a young age: by the age of fifteen, he was already active in the Bolshevik party under Lenin, and at sixteen he took part in the Russian Revolution of 1905.† After being found operating an illegal printing press a year later, he had been exiled to Siberia, from where, *I kid you not*, he *walked* to China, hopped over to Japan, and from there boarded a boat to San Francisco. Once in the US, he founded

---

* In other words, and to give you an idea of how much he packed into his life, you should read this chapter in the knowledge that we're completely leaving out the first two-thirds of it.

† This is known as the 'first Russian Revolution', and is usually overshadowed by the 1917 Revolution on account of not achieving any of the big showy things like getting rid of the Tsar, establishing a new form of government, etc.

the Communist Party of the United States of America and started working for Soviet intelligence.

Hemingway knew this, apparently, when Golos approached him with the offer of doing intelligence work for the NKVD. They had been interested in the writer for five years already, ever since he had written an article for the Marxist magazine *New Masses* lambasting the US government for their role in the deaths of more than 250 First World War veterans – men who had been sent, impoverished, unemployed and frequently suffering from what we'd now call post-traumatic stress disorder, down to work camps in what was then a remote and bankrupt corner of Florida known as the Keys.

In 1935, the Great Labor Day Hurricane hit, and it did so with an intensity that has yet to be matched by any other Atlantic hurricane. Hundreds died, and many more were injured or sickened in the aftermath of the disaster, and while Hemingway didn't go so far as to suggest that the government had *purposefully* sent the men to die in a hurricane, he certainly did imply that they weren't too bothered that it happened.

Then, in Spain, he had further proven his leftist credentials, growing close with Soviet and communist guerrilla fighters as they resisted Franco's insurgency – experiences that had very recently been released to the world, albeit in fictionalised form, in his new novel *For Whom the Bell Tolls*. That work would very soon be the centre of a literary controversy after the Pulitzer Board voted unanimously for it to win the 1941 Prize. Nicholas Murray Butler, the president of Columbia University and therefore de facto boss of the Pulitzer board, immediately vetoed this decision, on the grounds that the book was 'offensive', and while the standard explanation is that this was an objection to the sex in the novel, I'll just also point out that Butler was a big fan of the regimes of Mussolini and Hitler, and let you draw your own conclusions on the matter.

But if the Soviets thought all that would be enough to make Hemingway work for them, then … well, they were right, apparently. Golos reported back to Moscow that he was 'sure that he will cooperate with us and ... do everything he can' to help the NKVD.

Now, I don't want you imagining Ernest Hemingway, drunk celebrity author, trying to engage in some shady cloak-and-dagger tomfoolery with secret codewords and peep-holes cut out of newspapers. The Soviets knew he wasn't a spy. They also knew he was well-respected, and friendly with powerful people around the world – he even kind of had an in with the President since marrying Gellhorn, as she was friends with First Lady Eleanor Roosevelt. Sure, he might not have had access to secret documents or intel, the Russians thought, but he might prove valuable in other ways: maybe he could write stories with a subtly pro-USSR slant, or even just make introductions with other potential assets.

This worked out about as well as you'd expect, given it hinged on the cooperation of a man who once wrote a deliberately shitty book and destroyed a perfectly good friendship rather than just deliver on a contract.* Despite Soviet agents reporting that the writer 'repeatedly expressed his desire and willingness to help us', by the start of the 1950s they were pretty much cutting their losses: Hemingway had failed to ever 'give us any political information' and, they noted in diplomatic terms, was never 'verified in practical work'. In other words: useless.

Please don't think, though, that Hemingway did nothing good at all. The period between his recruitment and dismissal saw the start, and subsequent intensification, of the Cold War: by the end of the 1940s, the USA had gone full Red Scare, the USSR was testing atomic

---

* If you want to read a masterpiece in pettiness, it's called *The Torrents of Spring*, and it was probably – Hemingway denied it, but of course he would – written to get out of a contract with his first publishers. The writing is deliberately repetitive and antiquated, and Hemingway claimed he wrote it in a week. As a taster of the result: *'In some ways it was the happiest year of his life. In other ways it was a nightmare. A hideous nightmare. In the end he grew to like it. In other ways he hated it. Before he knew it, a year had passed. He was still collaring pistons. But what strange things had happened in that year. Often he wondered about them.'*

bombs in Kazakhstan, and it seemed like the world could only grow more divided. And yet, among all this, there was one thing that both global superpowers agreed on: Hemingway's ineptitude at spycraft.

In October 1942, about a year into the US's entry into the Second World War, a young FBI legal attaché named Raymond Leddy sent a letter to Director J. Edgar Hoover from his station in Cuba. He explained that 'a relationship ... has developed under the direction of the Ambassador with Mr ERNEST HEMINGWAY', and that he had been asked by the Embassy whether the writer would make a good intelligence asset. Despite Leddy's misgivings about the recruitment – after all, Hemingway had openly criticised the Bureau for arresting people enlisting to fight in Spain, and even worse, he had personally insulted Leddy at a jai alai match – the Embassy secretary approached Hemingway, asking him more or less: 'Hey Papa* ... listen ... hypothetically, if somebody asked you to, say, spy for the US government, in theory, for some unspecified reason ... what would you say?'

Leddy may have been holding a grudge against Hemingway for the comment at the jai alai match, but his doubts about the potential recruit were pretty swiftly proven correct. Hemingway's answer to the Embassy was a resounding 'yes – and also how about I set up my own marine spy squad that you pay for?'

You have to admire the balls of the man. Not only was he happily offering to be a spy for the mortal enemies of the guys he was *already* a spy for – oh, sure, they were both fighting on the same side at the time, but President Truman wasn't above openly hoping that as many Soviets died fighting the Nazis as possible – but he was also apparently the kind of guy whose *immediate reaction* upon becoming a double agent to both of the soon-to-be nuclear world powers was to hatch a most likely drunken plan to become a bona fide pirate of the Caribbean. And yet incredibly – *incredibly* – the Embassy said yes. And thus began

---

* Everyone called Hemingway 'Papa'. This is not made up.

Ernest Hemingway May Have Been the Worst Double Agent Ever • 279

Hemingway's 'Crook Factory': a ragtag band of local bartenders, sex workers, Loyalist Spanish refugees, Basque priests, fishermen, smugglers, and whoever else caught Hemingway's eye and imagination. This motley crew – or as the FBI termed them, 'an amateur information service' – would perform vital counterintelligence work against any Nazi spies hoping to infiltrate Cuba, Hemingway promised, thereby safeguarding one of the most important US allies in the Caribbean for the low low price of $500 a month.*

The Ambassador was delighted. Leddy was ... not. In fact, he was 'quite concerned with respect to Mr Hemingway's activities', the FBI noted, warning his superiors that 'they are undoubtedly going to be very embarrassing unless something is done to put a stop to them'.

You can almost feel Leddy's frustration through the page when you read the now-unclassified records. First you see him rebuffed by the Ambassador, who told him in no uncertain terms that any information brought to him by Hemingway must be acted upon immediately, and he's forced to report 'that Hemingway's information is valueless; that our Agents in Cuba have, of course, to check on it when it is submitted; that it is completely unreliable information; that the time taken to investigate it and check on it is purely wasted time and wasted effort; that Hemingway has not actually interfered with any investigation ... to date, but that from the way he is branching out with his undercover informants, he undoubtedly will.'

But if Leddy disapproved of Hemingway's spy gang antics, he must have *hated* Hemingway's other contribution to the War Effort: the *Pilar*.

What was the *Pilar*, you ask? It was, of course, a 38-foot fishing boat. But it could be so much *more*, Hemingway said, and so, with (somehow) the blessing of the Embassy,† he turned it into a one-man anti-submarine patrol vehicle.

---

* Around $8,500 per month in today's money. What a steal!
† Who by this point are really starting to embody the phrase 'play stupid games, win stupid prizes'.

'On the pretext of continuing [scientific] investigations, the Naval Attaché has acceded to HEMINGWAY'S request for authorization to patrol certain areas where submarine activity has been reported,' noted Leddy, somehow endeavouring to sigh through a typewriter. 'Special permits have been secured for this, and an allotment of gasoline is now being obtained for his use. He has requested that some firearms and depth charges be furnished him, which is also being done.'

Hemingway's plan, basically, was to sail around the Cuban waters, ostensibly fishing but actually watching for German U-Boats. Should he see one, he would wave them over for fish or fresh water or whatever, and then – bam! He and his crew would hit 'em with the machine guns. He even had a plan to recruit jai alai players to precision-lob grenades down the canning towers of the subs, a tactic which, if nothing else, was certainly original.

Unfortunately for those of us interested in the ballistic potential of weaponised jai alai players, Hemingway never fought a U-Boat – he reported seeing one, once, but it drove away before he got a chance to bazooka it. So he stuck to what he knew best: getting drunk and fishing. Keen to at least grenade *something*, he would occasionally bomb the fish rather than use a rod, but that was about the extent of the action he saw.

By April 1943, the US government had terminated any agreement it had with Hemingway regarding intelligence operations in Cuba, placing his Absolute Zero level tactical nous alongside the atom bomb in terms of things the US realised before the USSR. The Crook Factory was officially dissolved, and Hemingway went over to Europe to get divorced and break international law.

In June 1944, Allied forces landed at Normandy. Hemingway, on assignment for *Collier's* magazine as a war correspondent, was among them* – though he was kept on a raft at sea, away from the action.

---

* Making him technically one of the background characters in the opening sequence of *Saving Private Ryan*.

Ernest Hemingway May Have Been the Worst Double Agent Ever • 281

A couple of months later, he would have a final, small brush with intelligence fieldwork.

As Allied forces surged through France, on the eve of the Liberation of Paris, Hemingway found himself embedded with a band of partisan troops on the outskirts of the city. Overcome with the excitement of one more chance to punch a Nazi or two, he took off his press credentials, declared himself a colonel, and started directing the troops, leading patrols, gathering intel, and storing munitions in his quarters.

Which would have been nothing more than a fun epilogue if it wasn't for one pesky little *tiny* thing called the Geneva Conventions.*
See, under international law, it's what experts call 'a very big boo-boo' for journalists to take part in warfare – and Hemingway pretty soon found himself brought up in front of the Inspector General of the Third Army to answer for himself.

Hemingway faced his charges with all the honesty and integrity of a twelve-year-old caught with a cigarette. The mini-arsenal in his rooms weren't *his*, he protested, they were his friends' – he was just *storing* them. *Of course* he had military maps, what kind of idiot would visit a foreign country without a map?? And OK, maybe he'd taken off his press tags, but only because he was hot. It *was* August, after all. And being called colonel? Why, that was 'in the same way that citizens of the state of Kentucky are sometimes addressed as colonel', explained the Illinois-born Cuba and Idaho resident, 'without it implying any military rank.'

But the real *coup de grâce* of shamelessness came in answer to the big question: had Hemingway 'fought with the men?'

That's a pretty direct question. Not much wiggle room, you might think. Unless, that is, you feign a complete inability to parse how words work – which is a particularly bold strategy when you are famous *precisely because* you understand how words work.

---

* This was pre the big Geneva Conventions that we're used to now, but those are basically an edited best of collection of a bunch of Geneva Conventions that existed beforehand.

'I swore I did not "fight with the men",' Hemingway would explain years later. 'Who would fight with the men, i.e. not get along with them. Not me ... I denied and kidded out of all of it and swore away everything I ever felt any pride in.'

In 1945, Hemingway returned to Cuba, where he stayed for a decade and a half, entertaining famous guests, occasionally travelling, writing, and drinking himself into an early grave. In 1953, twelve years after his initial snubbing, he was finally awarded the Pulitzer Prize; the next year he won the Nobel Prize for Literature. Through it all, the FBI watched him.

And then, at sixty-one, he died. It happened at his family home in Ketchum, Idaho, where he had been living for less than a year, leaving behind scores of awards and admirers, a half-finished book with a deadline only weeks away, and a legacy that proved as mysterious to his friends as to his fans.

'To have known him, at least to have known him superficially and late in his life, makes it more rather than less difficult to understand him,' wrote Robert Manning, the editor of the *Atlantic*, who had spent time interviewing Hemingway in 1954. He described his friend as a 'big man', barefoot and wearing sloppy khaki shorts, built like a linebacker and knocking back martinis as he recounted stories from his life and offered up his philosophies on writing and the world.

Ernest Hemingway was an incredible writer, but he was a godawful spy – which makes it all the more baffling that he opted to spend so long embracing the vocation twice over. Honestly, I've nothing but respect for the guy: by the end of his life, his health had been wrecked by years of alcoholism and paranoia, he was being stalked by the feds, and he'd spent decades in self-imposed exile from his home country, ridden with anxiety that his history working for the Soviets might find him in front of one of the new McCarthyist 'committees' – and all that for a job that nobody forced him to do, and in fact he was so laughably bad at that it almost seems like he was simply trolling these two global nuclear superpowers for his own personal amusement.

After all, as he wrote one Christmas to Manning: 'We had fun, didn't we?'

29.

Yukio Mishima and the Shortest, Gayest Fascist Coup in History

'The beautiful should die young, and everyone else should live as long as possible,' wrote the thirty-one-year-old author, poet and playwright Yukio Mishima after James Dean's fatal crash on California's Route 466. And Yukio Mishima was beautiful.

There was really no way that Kimitake Hiraoka, the child who would grow up to become Yukio Mishima, could avoid dying in an ill-fated nationalist coup. He was primed for it from birth: at less than a month old he was more or less kidnapped by his overbearing grandmother, an illegitimate daughter of a daimyo,* who had been raised to be the wife of a samurai ruler and resented the fact that she had become the wife of a bureaucrat instead.

---

* The daimyos were feudal lords who collectively ruled over most of Japan for about 900 years – though by the time Kimitake was born, they were at best figureheads rather than any kind of actual political force. Basically, imagine if your grandma was the illegitimate daughter of the Lord Mayor of London – you know, probably pretty posh, and very braggable if you happen to be in London and a Tudor, but, if you're honest, these days nobody really knows who they are or what they do, and now most of your gran's time is spent dressing up in ceremonial garb from the old days and wishing the non-lord Mayor of London would know his bloody place.

And by 'overbearing', we're not talking about, like, a grandma who spends too much time nitpicking over parenting style or gets way too invested in baby dance class or whatever. This is more like the grandma who steals your baby and then doesn't let him go outside or see other boys or the literal sun for more than a decade style of overbearing. She even restricted the amount Kimitake's mother was allowed to see her infant son: only for scheduled, time-limited, breastfeeding sessions. Kimitake's earliest memories would have been of his grandmother instilling in him a reverence for samurai lore and tradition; she ruled over him with a flair for the dramatic that sometimes veered into violent outbursts, and he spent his formative years basically confined to her bedroom until he was reunited with his parents at age twelve.

That sounds pretty bad, but it's hard to say things got better once Kimitake got back home. Like many children forced to live in one room with only a grandma for company, he had returned to his parents a frail, shy tween who was interested in literature and beautiful broken men, and his father *hated* it. Toughening up was required, and to Hiraoka senior that meant military discipline and parenting tactics like ripping up his son's favourite books, holding him up to the side of a speeding train in an apparent attempt to scare him straight and poisoning his cat.*

Fortunately for Kimitake, his father's reign was short-lived – he got a promotion shortly after Kimitake came home and moved, alone, to his new job in Osaka. Baby gay Kimitake, consummate mama's boy,† was finally free to indulge in all the literary pursuits and pleasures his father had denied him: he wrote hundreds of poems, haiku and tanka, and his works appeared alongside those of much

---

* The cat survived, you can relax.
† To a kind of gross degree, actually. There are stories from friends of the family that describe his relationship with his mother as 'incestuous' – one example saw Kimitake literally licking his mother's feet to soothe them after she complained of them hurting.

older boys in the school magazine. This isn't an idle boast, by the way: Kimitake attended the Gakushūin, or 'Peers School'. As well as educating the sons and daughters of emperors, the Gakushūin was also notable for having produced a highly influential literary circle in the early twentieth century known as the Shirakaba-ha – literally, the 'white birch society'. It would be like getting published in your school magazine if your school also happened to be where the entire Bloomsbury set attended.

So by the time he was sixteen, Kimitake was the culmination of years of isolation and indoctrination around Japan's glorious feudal history and his rightful place at the top of it, which had been further honed by a militaristic father whose parenting style would have been considered 'a bit much, actually' by the CIA guys at Guantanamo. The various hang-ups this understandably left him with were deepened further during the Second World War, when Kimitake was declared too sickly to fight for the Emperor and had to work in a factory instead – hardly the work of a samurai, and extremely unlikely to grant him the glorious young bushidō death that by this point had been baked into his psyche.

But something else happened to Kimitake during the war: mere months after Japan entered the fray, his first short story, *'Hanazakari no Mori'* ('Forest in Full Bloom'), was published in a prestigious conservative literary magazine. It had been hailed as a work of genius by his teacher and the editorial board, who all congratulated themselves for finding such a wunderkind. It could have been a dream come true for the young Kimitake, if not for one thing: his father's hatred of all things literary.

And so a solution presented itself. Kimitake Hiraoka would stop dreaming of becoming a writer, would support Japan in the war with factory work and then follow his father into a government job. He would marry Yōko Sugiyama, a very understanding woman who, in his words, 'was the daughter of an artist, so … wouldn't hold too many of the illusions people have about artists' – which is a red flag –

and would abide by his demands that she keep herself separate from his private life, did not 'interfere' with his writing or bodybuilding, and generally be shorter than him,* which is enough red flags to start a communist uprising. He would father children.

And Yukio Mishima, his alter-ego named after the snow on mount Fuji and a fortuitous train station on two of the editors' commute, would become one of the most important novelists in modern Japanese history.

Of course, as I'm sure anybody who has even contemplated living a double life on this scale would expect, this split did not last. An office job was never for Kimitake, and, frankly, marrying any woman at all is a bit of an odd move for a young thirty-something man known for frequenting gay bars and writing multiple books about fancying men. And, yes, I'm aware bisexual people exist, and it's by no means impossible that he was genuinely attracted to his wife, but given how he treated her, it's perhaps more likely that this marriage was yet another double life, worn for the sake of propriety by a man hungry for honour in a time and place where being unmarried past thirty was suspicious and his dying mother was imploring him to settle down with a nice girl and make some grandbabies.

And so, as recorded in his father's memoir, *My Son: Yukio Mishima*, the young Kimitake announced one rainy night that he was quitting his respectable day job to become a writer. And, almost immediately, he became a superstar.

Mishima's second novel† was titled *Confessions of a Mask*, and it told the completely made-up, totally fictitious, not based in reality at all story of a frail, shy boy forced to spend his childhood confined to his grandmother's bedroom, and the evolution of his awakening as a

---

* Since you're now wondering, he was at most about 5 foot 3 inches.
† His first novel, *Tōzoku* (*Thieves*), a love story about a pair of suicidal aristocrats, had been released in serial form a couple of years beforehand, but it kind of bombed and these days everybody just kind of pretends it never happened. It's not been translated, and it's rarely seen even inside Japan.

gay young man in contemporary Japan. Also the main character may or may not have been named Kimitake.*

Now, there's a long literary tradition in Japan known as shi-shōsetsu, which Mishima thought was pretty worthless: he considered it self-indulgent, unimaginative and without talent. It translates to 'I-novel', and it basically refers to an autobiography presented as fiction. If that strikes you as an incredibly good description of *Confessions of a Mask*, that's because it is – though Mishima himself would later call it 'confidential criticism' and claim he had personally invented the genre.

So aside from being Definitely Not One Of Those Garbage Autobiographical Novels, *Confessions* can be used as a lens through which to understand Mishima throughout his life. It showcases his obsession with beauty and sex and death; his self-consciousness about his own body and his desire for those of strong, masculine men. Most of all, it explicitly blurs the lines between real life and performance: respectability, as Mishima saw it, was a 'reluctant masquerade' that he was doomed to play out.

Throughout his twenties, the image he had cultivated for himself was one of an aesthete and a dandy: he wore high-end, Western-style suits and referenced European literature and philosophy; photos from the time show a delicately-featured man smoking a cigarette at his desk or playing with his cat. His next novels continued the scandalous themes of *Confessions*: he confessed more of his sexual secrets in *Forbidden Colours*,† a novel about a young gay man being

---

* The protagonist is named Kochan, which is basically the equivalent of somebody named Robert writing a book where the main character is called Bobby.

† *Forbidden Colours* was written during a period of Mishima's life when he was known to frequent the gay scene in Tokyo. The original Japanese title, *Kinjiki*, is written using two Japanese *kanji* pictographs, which can, and in some contexts absolutely does, translate to 'forbidden colours' – in the context of the novel, however, there's a second meaning to the *kanji* that gets lost in translation: forbidden erotic love.

manipulated by an aging and cynical writer; in *Temple of the Golden Pavilion*, he wrote about the burning of the 600-year-old Kinkaku-ji temple in Kyoto – Mishima, true to form, imagined the destruction of the temple to be the acme of its beauty.

But as he grew older, the young man who longed to die like James Dean began to fade. Instead, Mishima's insecurities about his body and his perceived weakness took over: he started a strict body-building regimen and took up karate and traditional Japanese sword fighting. His flair for the dramatic remained: he modelled;* he was a movie star, playing a repentant yakuza in 1960's *Karakkaze Yarō* ('Afraid to Die') and an elite Samurai warrior in 1969's *Hitokiri*, as well as a handful of extra roles in the dozens of film adaptations he brought to screen. He started to shun philosophy and words in favour of physicality, longing to become a 'man of action'.

'I have believed that knowing without acting is not sufficiently knowing,' he wrote in a letter to a friend towards the end of his life. 'After thinking and thinking through four years, I came to wish to sacrifice myself for the old, beautiful tradition of Japan, which is disappearing very quickly day by day.'

Perhaps you can already see where this is going. Mishima had always leaned to the right, politically – having a childhood spent in isolation with an old woman obsessed with feudal honour systems can do that to a boy – but as he moved into middle age, things started hurtling into fascism and neonationalism very quickly, and in quintessentially Mishima fashion.

---

* For all Mishima's griping about how self-indulgent shi-shōsetsu could be, the man sure did like to indulge himself. One of the most iconic shots of the writer has him strung up against a tree, nearly nude and seemingly shot through with arrows in an homage to Saint Sebastian, and thanks to a rather personal admission in *Confessions of a Mask*, we can be pretty sure that the staging of this photo was no coincidence. Mishima was, to put it bluntly, *extremely* horny for Saint Sebastian – and let's face it, a young, muscular, barely-clothed Adonis on the cusp of death, as the third-century martyr is generally and most famously portrayed, is incredibly on-brand for Mishima.

In 1968, at the age of forty-three, Mishima founded the Tatenokai, or 'Shield Society'. There were about 100 members, most of whom were young men from the local university. The society was meant to be an anti-communist defence force – the group was even permitted to train with Japan's national armed forces – but even from its conception, there were rumours in Japanese society that Mishima had set it up as a sort of fascist prototype Grindr. It's kind of easy to see why: a group of young athletic men devoted to working out – Mishima oversaw their training regimen himself – wearing fancy expensive clothes bought for them by Mishima, and all of them ready to die for the honour of their Emperor and Mishima.

Here's the thing though: by the late 1960s, swearing your life to the Emperor was a bit weird. Sure, he technically was still officially 'the symbol of the State and of the unity of the people', as per the national constitution, and, yes, he is responsible for convening and dissolving the chambers of government – but other than that, his job had been entirely ceremonial ever since the end of the Second World War. He wasn't head of the army – not even in theory, like UK monarchs get to be – and he certainly wasn't officially divine any more; although he retained the title of Tennō, or 'heavenly sovereign', he had formally and explicitly rejected the idea that he was a direct descendant of the sun goddess Amaterasu* in an imperial decree of 1946.

Despite all this, on 25 November 1970, a battalion stormed the headquarters of the Eastern Self-Defence Force of Japan. The battalion in question consisted entirely of Yukio Mishima, his probable boyfriend Masakatsu Morita, and three other members of the Tatenokai. None of the boys were older than twenty-five years old.

What happened next has gone down in history as one of the most bizarre and confusing coup attempts of all time. The Shield Society members took the commander of the Japanese Defence Force 32nd

---

* This kind of thing has historically been a neat way to get out of things like parking fines or armed rebellions. Try it for yourself.

regiment hostage for more than two hours while Mishima addressed the soldiers below. Dressed in a brown uniform and wearing a samurai headband on his head, he shouted at the troops to reject the post-war constitution that had renounced Japan's armaments and rendered the Emperor human; to join his uprising and save Japanese culture. He had planned to speak for thirty minutes.

He made it to about seven. For all the months of planning that had gone into this insurrection attempt, Mishima had failed to take into account one possibility: that the soldiers would say no. As he harangued them for having lost their samurai spirit, they jeered and heckled him, telling him to calm down and stop acting the hero. Much of what he managed to get out was drowned out by nearby helicopters. He gave up, having converted precisely zero soldiers to his cause, and retreated back into the commander's office.

'I had no choice but to do this,' he told the hostage commander. Then he took his sword, and, like the samurai he knew he has born to be, he disembowelled himself in the ritual suicide of seppuku.

Morita had been designated his second for the seppuku – a position that gave him the important job of decapitating Mishima to save him unnecessary pain. Just like the rest of the coup, this was a slapstick failure, and after three failed attempts to cut off his alleged lover's head, another one of the group stepped in to finish the job.

The revolt and suicide of one of Japan's leading lights of literature did not revolutionise Japanese society like Mishima had hoped. Instead, the country – and the world – reacted more with bafflement. His political allies derided him as a madman; others called him a showman who had designed his suicide just like he would design his novels. Still others claimed it was an act of eroticism to the conflicted writer, or just Mishima's escape from growing old.

'Those who would have been his ruin, had he lived, are still distraught over his passing,' Mishima had written when Dean died. 'But is it really such a shame their hands never had a chance to tarnish him? Is it really such a shame that he was wise enough to make the first move, to take the lead and soar ahead, beyond the reaches of the snatching masses?

'The public is our marker for the cruel passage of time, and time is the perennial victor, but it will never shake the memory of this rare, happy, priceless stroke of defeat.'

30.

NASA Forgets about Women, Toilets and the Metric System

Here's an introduction you probably weren't expecting for a chapter about the history of NASA: let's talk about penises.

Now, it may be true in the world of romance that size isn't everything, but once you leave the planet, it really does matter. No, not in the way you're thinking of – research* suggests that sex in space would not actually be much fun. No, I know what you saw Captain Kirk getting up to with all those Orion slave girls on *Star Trek*, but trust me: thanks to the effects of bobbing about in zero-G, sex in space would most likely be hot, stinky, difficult to manoeuvre, and surrounded by floating blobs of sweat and puke. If you could cope with all that, there's still the fact that your blood pressure usually decreases in space, reducing the amount of blood flowing into the, ah, extremities of the body. And if you don't know what effect that would have on the prospect of some zero-gravity romance ... well, go ask your parents.

---

* *Of course* people have researched this.

And to be honest, even if it was spectacular we wouldn't know about it, as both NASA and Roscosmos have categorically stated that no astronauts have ever got it on in outer space – so either no humans have yet had the pleasure of, well, the pleasure, or there are some serious NDAs at work to keep us from knowing the X-rated truth.

But that just makes it all the more surprising when you hear that the second American to ever enter space did so wearing lingerie and a condom.

That's not a joke: according to popular legend, Gus Grissom, decorated two-time war veteran and Air Force pilot, spent about quarter of an hour in July 1961 off-planet in a garter belt and a safe sex aid.* Each to their own, of course, but it doesn't sound like the most practical thing to wear under a spacesuit – unless, that is, you're comparing it to what his predecessor had to wear, which was a spacesuit full of his own urine.

You'll be relieved† to know that 'covered in piss' wasn't how NASA had actually planned to send the first American into space. It wasn't how astronaut Alan Shepard (or really any of us, I imagine) wanted to be remembered in history books either, which is probably why, once nature called, he very sensibly radioed down to ground control to ask whether he could go peepee before blast-off.

Unfortunately, NASA had been so wrapped up with trying to beat the commies in the Space Race that they had apparently plum forgot that humans sometimes need to pee. In other words: there was no toilet on board. If Shepard wanted to relieve himself, he would need to exit the spacecraft, hope like hell that someone had left a Portaloo next to the launchpad for some reason, and get back in time for blast-off, and that meant a whole hours-long process of depressurising the air inside before he could leave and repressurising it once

---

* Other stuff too, of course, but that's less funny and therefore why mention it.
† See what I did there?

he got back. Not inclined to go through all that just for the sake of Shepard's bladder, ground control told him instead to, well, just be a big boy and hold it in a little longer.

Of course, it doesn't take a wild amount of deductive reasoning to realise that a decorated naval aviator and test pilot wasn't going to risk screwing up his chance to be the first ever capitalist in space if it wasn't, you know, urgent. In an exchange that has for some reason been removed from the official transcript, Shepard informed ground control that he was going to urinate, and they could either let him out of the expensive, cutting-edge, and notably un-peed-upon spacecraft to do it – or not.

But, like the too-stubborn supply PE teacher adamant that you've forgotten your kit just to mess with them, ground control refused to relent. So Shepard did the only thing he could: he gave enough notice for control to shut off the biomedical sensors that covered his body and spacesuit, and then, as heroically as possible, he wet his pants. And because, while you're in a rocket waiting for blast-off, you generally have to be on your back, this left the Free World's first great pioneer in the Space Race lying face up in a puddle of his own wee.

There was a real fear at the time that the urine might short-circuit the electronics in Shepard's suit and electrocute him to death, too, so if you were wondering what the metaphorical equivalent of having to do PE in your vest and pants was, it's apparently 'pissing yourself to death while lying face up in a cramped capsule'.

After drying off, Shepard spent about fifteen stinky minutes in sub-orbital space before hurtling down into the Atlantic Ocean, where he and the space capsule were retrieved by a recovery helicopter. And to be fair to NASA at this point, it's easy to see why they screwed up here: the reason they never accounted for a full bladder was that they figured an astronaut ought to be able to hold it for *quarter of an hour*. They didn't really forget that humans wee, they just forgot that rocket launches take hours – eight hours altogether, for Shepard – regardless of how long the spaceflight itself ends up being.*

Still, the West had finally joined the USSR in slipping the surly bonds of Earth, and in doing so had learned an important lesson: humans need to pee.

And that's where Gus Grissom and his kinky spacesuit comes in. Now, the second manned US spaceflight had modest aims, being more or less a copy of the first, but it differed in one crucial detail: this time, the astronaut would hopefully not leave Earth while smelling of wee. Of course, they couldn't just cram a toilet into the existing shuttle (I mean what plumber has insurance to cover that?) and so, the story goes, an incredibly resourceful mission nurse named Dee O'Hara decided to improvise: she popped to the shops, bought a condom and a garter belt, and after a few minutes' crafting presented Grissom with a makeshift pee container.†

I wish I could tell you that NASA decided to create some kind of awesome space-toilet after this, but unfortunately for both us and subsequent space-peers, they decided instead to employ the celebrated engineering principle of 'If It Ain't Broke, Don't Fix It'. Basically, they formalised the get-up innovated by Nurse O'Hara, and designed the UCD, a device which sounds impressive until you learn it literally stands for 'urine collection device' and was just another condom-with-a-tube-attached contraption but a bit fancier and way more expensive.

Now, if you're wondering how well these UCDs worked at pee collection, I invite you to reread the description 'condom with a

---

* Then again, to be even more fair, they're NASA, and if anybody knows how long a rocket launch can take, it really ought to be them.

† Fun though this story is, Grissom actually solved the pee problem by wearing two pairs of rubber trousers, which, you might argue, is not less kinky than a garter belt. He urinated in between the pairs, leaving him orbiting the Earth as the bottom slice in an astronaut-wee-spacesuit sandwich.

'[There is a need to] make more rigid demands on urine collection,' NASA scientists admitted in a post-mission press conference.

tube attached'. The devices leaked so much that test subjects refused to wear them, presumably figuring that if they were going to end up covered in piss anyway they may as well be comfortable while it happened.

Thus began one of the most R-rated research and development projects in spaceflight history – and it was undertaken by just one man. In 1961, NASA hired James McBarron to head up its pee division, and it's fair if slightly disturbing from a visual metaphor point of view to say he threw himself into the job. He bought packs of every brand of condom he could find, which I imagine is a bit less embarrassing when you can tell the judgemental cashier: 'it's OK, I work for NASA'. He used himself as a test subject, pissing into condom after condom trying to work out which would make the best, most watertight UCD. Finally, just in time for John Glenn's spaceflight in 1962, NASA reported that 'A satisfactory UCD has been developed … for immediate flight requirements.'

The new system worked well. Glenn went down in history as the first person to do a wee in orbit, emptying 800ml of urine – which, for the record, is *a lot* of urine; without gravity to tell you your bladder is filling up, you don't start to feel the urge to go until much later than here on Earth – into his fancy new collection device. Problem solved. The only thing the NASA scientists hadn't taken into account was male ego.

Obviously, all men aren't created the same. There's nothing wrong with that; it's quite literally why clothes come in different sizes. Clothes and, of course, other things too.

So when NASA started equipping their astronauts with the new UCDs, they understandably offered a range. The astronauts were tasked with choosing the appropriate size out of Small, Medium or Large, and I'm guessing you're already starting to see where this is going.

'[T]here's always this little ego thing about which one you do pick,' confessed astronaut Rusty Schweickart in a 1976 interview. 'Of course the smart guy picks the right size, because it's very important. But … [if] you've got an ego problem and you decide on a large when you should have a medium, what happens is you take your first leak and you end up with half of the urine outside the bag on you. And that's the last time you make that mistake.'

Schweickart might have been the one to admit it, but penis anxiety was evidently enough of a widespread problem that NASA had to change certain things about how they operated.

'[F]ew astronauts, whatever their real dimensions, refused to accept that they were anything but large,' explained zero-gravity waste management guru Donald Rethke (also known as, and I am not kidding here, 'Dr Flush'). 'We changed the names to large, gigantic, and humongous.'

The answer to this childish behaviour was both simple and pleasingly poetic. With the advent of female astronauts in 1978, NASA tried to come up with a women's equivalent to the UCD – but saw little success. It wasn't until the eighties that they hit upon an idea that had been originally invented a mere few millennia beforehand: nappies.

It's kind of amazing it took them so long to come up with this, since they had already diapered their male astronauts for long-haul flights: Neil Armstrong's historic One Small Step was made in a nappy, and while Buzz Aldrin may not have been the first man to walk on the Moon, he claimed his entire life to have been the first person to take a leak there.* These nappies, or 'Disposable Absorption Containment Trunks', as NASA preferred to call them, were specially designed to be comfortable and leak-proof, making the new female recruits the envy of the space-faring crowd. Then, in 1988, somebody at NASA finally learned what anybody who's walked down the feminine hygiene aisle in a supermarket already knows: that adult diapers

---

* We all play the hands we're dealt, I suppose. Incidentally – a fun fact that you probably didn't read in history class is that there are ninety-six bags of human poop on the Moon, thanks to Armstrong and his compatriots.

exist. You can just buy them, it's really easy. They've literally been there the whole time, NASA.

Of course, if anybody at the agency ever *had* walked down that aisle, then they probably wouldn't have made another of their more infamous screw-ups – another bathroom-based mistake, as it happens. Now, as you may already be aware – and if you aren't, then I suggest you put this book down and pick up something more age-appropriate before somebody tells you off for reading swears – there's approximately half the planet out there who are forced to have a period every month. Not all at the same time, of course (that would be terrifying), and not all in the same way – they can be long or short, heavy or light, cripplingly painful or barely noticeable. But overall, the uterus's regular-ish ritual of suddenly noticing that there's no baby in here *again*, Susan, we *talked* about this, and throwing a three-to-seven-day-long tantrum to teach you to be *less responsible* in future, *Susan*, is one that humankind has known about since, well, probably before 'humankind' was even a thing.

So menstruation may be annoying, but it's not rocket science. Which is perhaps why NASA, an organisation pretty famously crammed full of rocket scientists, managed to screw up so badly when faced with the very normal and everyday task of, um, buying some tampons.

The June 1983 Space Shuttle mission STS-7 was the seventh that NASA had launched in a little over two years. The agency had been hurling people into space for more than two decades by that point, and it wouldn't have been surprising if the miracle of human spaceflight was becoming a little, well, routine.

Except. You see, if we're being specific (and if there's one thing we'll be learning about spaceflight today, it's that it's good to be specific), NASA hadn't been sending people into space – they'd been sending *men*. And STS-7, the first mission to include members of the new Group 8 astronaut class in its crew, was going to make history – because STS-7's crew included a woman.

Now, to give you an idea of just how uniform the previous classes of astronauts had been, Group 8, on its own, included the first ever American woman in space, the first ever African-American in space, the first ever Jewish-American in space, the first ever mother in space, the first ever Asian-American in space, the first ever African-American to pilot and command a mission, the first ever American woman to perform a spacewalk, and the first ever American woman to make a long-duration spaceflight – and *these were all different people*.* But the problem with that was that – despite having put men on the Moon, sent frogs into orbit, and kept mice on a space station – when the time finally came to deal with a handful of highly-trained and eminently qualified human women, NASA found itself truly flummoxed.

'The technical staff at JSC [NASA's spaceflight training centre] – around 4,000 engineers and scientists – was almost entirely male,' recalled Sally Ride, the astronaut who was about to make history by having a vagina in space. 'There was just a very small handful of female scientists and engineers – I think only five or six ... the arrival of the female astronauts suddenly doubled the number of technical women at JSC!'

But just because NASA was finally willing to send women into space didn't mean they were ready to, you know, listen to their opinions. Which is how, in 1983, NASA ended up deciding that for her six-day mission, Ride would need about 100 tampons.

Now, it's true that every period is different, but even those of us with an exceptionally heavy flow rarely have to change tampons every eighty-seven minutes, night and day, for six days straight. That's hospitalisation territory. If Ride was getting through 100 tampons in less than a week, you could forget about all the ridiculous concerns floated by misogynists of the time about her screwing up due to PMS – she'd be more likely to be passed out in the cockpit due to losing about 5 per cent of her blood volume.

'Is 100 the right number?' NASA asked Ride.

---

* It also included the first LGBT astronaut to go to space – although NASA and the public didn't know it at the time.

'That,' Ride replied with what must have been a superhuman level of self-control, 'would not be the right number.'

It's a good thing they asked in the end, since it would have been at least five yards of tampons, all connected by the cords for safety like a particularly absorbent string of sausages. I don't know if you've seen *Gravity*, but there's not much space up there for what is, to use the SI system of metric prefixes, a full hecto-tampon – and that's if NASA didn't screw it up even further, since, apparently, double checking their working was not their strong suit.

See, we're used to thinking of the agency today as being pretty much at the forefront of human knowledge, bringing us images of the furthest reaches of the universe and urging forward our understanding of existence. And throughout the Cold War – that is, up until the early nineties – they had been equally revered as the heroes of the West,* showing the world the benefits of capitalism, an economic system so fantastic that it could put a man in space mere weeks after the communists already did the same thing.

But, in the nineties, NASA's reputation was flagging a little. A new semi-official policy of 'faster, better, cheaper' and a string of costly and embarrassing failures had earned the agency the nickname 'Not Another Space Attempt' by December 1998, when they were about to launch the Mars Climate Orbiter. It was the second probe to be sent up in NASA's new Mars Surveyor programme, and its goals were simple: to study the Red Planet from orbit, and relay messages from surface probes back to Earth.

At first, everything looked good: at 9 a.m. on 23 September 1999, right on schedule, the MCO entered the orbit of Mars. The programme scientists back at NASA had calculated that the probe would stabilise into a near-circular, Sun-synchronous orbit by December, at which point it would be fully operational and ready to broadcast new insights into Earth's closest neighbour.

---

* Or the filthy capitalist pig-dogs, depending on which side of the Iron Curtain your allegiances lay.

But the orbiter would have to take the long route to success. To get from entering orbit to operational orbit, it would have to fly behind Mars, temporarily losing contact with mission control, before coming back on the other side of the planet roughly twenty minutes later.

At 9.05 a.m. – just a little earlier than anticipated – radio contact with the orbiter was lost. NASA waited.

And waited.

Two days later, the mission was declared a loss. The probe, NASA would eventually discover, had been heading over a hundred miles too close to the surface and had simply disintegrated under the extreme stresses of the Martian atmosphere. The orbiter, and the $125 million that it had cost,* was gone.

The failure of yet another mission might have been painfully embarrassing for NASA, had the reason behind it not been even more humiliating. Because the thing is, the loss of the Mars Climate Orbiter could have been avoided if the engineers had only compared notes at some point.

By 1999, most of the world had made the sensible decision to switch to metric units, especially in the sciences. The one major outlier, of course, was the USA, where they were still customarily using an ancient and nonsensical system of units like 'feet' and 'pounds'.† So when NASA, an agency crammed full of sciencey type people used to measuring force in Newtons, decided to use a piece of navigational software supplied by Lockheed Martin, an American corporation with a preference for measuring force in pounds, they really ought to have double checked their working before they sent hundreds of millions of dollars' worth of equipment up into space.

---

* Of course, this isn't the whole story. All told, counting development, equipment, launch, operations, the whole shebang, NASA reckons it lost about $327.6 million on this disaster.

† Yeah, I said it! Come at me, Imperialists; it's the twenty-first century, it's time to switch to a system that makes sense.

Basically, the probe was being run by a system that measured the effect of the orbiter's thrusters in imperial units and accounted for that effect in metric, without at any point converting between the two systems of measurement. Guided by readings that were, *at best*, incorrect by a factor of nearly four and a half, the probe had veered way off course, burning up in space after managing to take just a single photograph.

It was an embarrassing loss, but it could have been forgiven – if only they hadn't made the exact same mistake six years later.

Yes, incredible as it may seem, NASA failed to learn their lesson from this high-profile extra-terrestrial fuck-up, and, in 2005, NASA's DART mission – a project designed to test the ability of spacecraft to rendezvous and manoeuvre in close proximity without human guidance – ended less than halfway through its expected timespan after the test spacecraft crashed into the military satellite it was meant to dock with, unilaterally decided it was out of propellant, and settled into an early retirement.

It was a high-profile mission, with high risk and – had it worked – high reward. But unfortunately, it was most likely doomed from the start: a post-mortem of the mission found that the engineers were inexperienced and under-supervised, the deadlines were too short and the technology was unproven.

But one very *specific* mistake that probably could have been avoided was to have the spacecraft read the GPS data in feet after having measured it in metres.

This mismatch set off a chain of errors in the trajectory until eventually 'DART was flying toward [the satellite] at 1.5 metres per second while its navigational system thought it was 130 metres away and retreating at 0.3 metres per second,' explained the official investigation report.

It would be nice to say that after losing a $125-million probe on Mars then six years later losing a $110-million spacecraft in the exact same way, NASA learned its lesson regarding measuring things properly. Unfortunately, despite nominally converting to metric back in 1990, in practice the Agency continues to use a mix of metric and imperial to this day.

'Although use of SI in the U.S. is increasing, aerospace is recognized as one area where adoption will be difficult, due to the long-standing use of the US-based "inch-pound" system for aircraft,' a 2013 statement from the Office of the Chief Engineer declared.

Luckily, NASA has at least promised to keep the Moon an imperial-free zone.

'When we made the announcement at the meeting, the reps for the other space agencies all gave a little cheer,' reported NASA representative Jeff Volosin. 'I think NASA has been seen as maybe a bit stubborn by other space agencies in the past, so this was important as a gesture of our willingness to be cooperative when it comes to the Moon.'

Yeah, it could be that, Jeff – or it could just be relief that we don't have to watch the Moon accidentally get smashed into pieces because some engineer in Houston forgot how many gigantamax condoms per fortnight there are in a lightyear. Just a thought.

Epilogue

So, class, what have we learned? All good stories have a moral after all; we've just rollicked our way through thirty, so you'd hope there'd be at least *something* we could take away from the experience.

But as much as the figures in this book are linked by their 'genius' reputations, as individuals, they're as different as you and me. Is the secret to going down in history taking risks and rebelling against the system, like Marie Curie showed us? Maybe. Alternatively, that might be the dumbest thing you ever did. It might mean dying in the world's shortest, most confusing fascist coup, and being remembered more for being a weirdo right-wing mummy's boy than for being a literary pioneer.

Can we take from our stories some message about history being shaped by great and influential individuals? Yeah, sure, maybe. Unless you're trying to discover a new planet, of course, at which point being able to work well as a team really comes in handy. But not *too* well, or else you run the risk of losing a million-dollar spacecraft somewhere off the edge of Mars.

So what, if anything, should you take from all this? Well, it's not like me to quote a hedge fund investor, but allow me this one indulgence: there's something Warren Buffett once said, about the unpredictable role of luck in a person's life. Musing on his success, he reflected on what might be his greatest accomplishment in life: being born in a time and place that rewarded his particular skill set many times more than its objective worth.

And he was right. Can you eat investments? No. Do they bind wounds or care for the vulnerable? No. They're not even very entertaining, unless you happen to have an unusual soft spot for Brownian motion – and trust me, very few people outside of a niche area of stochastic mathematics do. Valuing businesses, Buffett admitted, is not 'the greatest talent in the world … It just happens to be something that pays off like crazy in this system.'

Most of us will never be that lucky. But the people in this book, I think, were. They may have been a bunch of idiots running around doing some of the stupidest shit imaginable, but they were also people just good enough at some hyper-specific talent – writing sonatas, or thinking up mathematical smackdowns, or … I dunno, creating cults? – that we've rewarded them throughout history with far more respect and awe than they arguably deserve.

To put it another way: Einstein may not have said that thing about the fish being judged by its ability to climb a tree or whatever, but whoever *did* was on to something. Maybe it just needs a little tweak. The truer proverb, I would say, is something like this: we're all idiots – but if you judge an unkempt, socially inept love rat with a potentially fatal love of sailing holidays on his ability to do relativistic physics, then the whole world will think he's a genius.

References

1. Pythagoras

Copleston, Frederick, *A History of Philosophy*, vol. 1 (Doubleday, 1993)

Ewbank, Anne, 'Why Beans Were an Ancient Emblem of Death', *Atlas Obscura*, 25 May 2018

Huffman, Carl, 'Pythagoreanism', Edward N. Zalta (ed.), *The Stanford Encyclopedia of Philosophy* (2019), https://plato.stanford.edu/archives/fall2019/entries/pythagoreanism/

Khafizova, Taliya, '"Musica Universalis" Theory by Pythagoras: Global and Human Perspective', Girne American University

Lienhard, John H., 'The Engines of Our Ingenuity: The Pythagoreans', University of Houston, www.uh.edu/engines/epi213.htm

O'Connor, J. J., and Robertson, E. F., 'Arabic mathematics: forgotten brilliance?', MacTutor History of Mathematics Archive, University of St Andrews, 1999

Pennington, Bruce, 'The Death of Pythagoras', *Philosophy Now*, issue 78: April/May 2010

2. Confucius

Adler, Joseph A., 'Confucianism in China Today', Kenyon College, Pearson Living Religions Forum, April 2011

Confucius, *The Analects*, translated by James Legge (1893)

Csikszentmihalyi, Mark, 'Confucius', Edward N. Zalta (ed.), *The Stanford Encyclopedia of Philosophy* (2020), https://plato.stanford.edu/archives/sum2020/entries/confucius/Dubs

Encyclopaedia Britannica, The Editors of, 'Spring and Autumn Period', *Encyclopedia Britannica*, 2017, www.britannica.com/event/Spring-and-Autumn-Period

Entwisle, Barbara, and Henderson, Gail E. (eds), *Re-Drawing Boundaries: Work, Households, and Gender in China* (University of California Press, 2000)
Goscha, Christopher, 'What is Confucianism?', *History Today*, 9 March 2017
Hendricks, Robert G., 'The Hero Pattern and the Life of Confucius', *Journal of Chinese Studies*, vol. 1, no. 3, 1984
Homer, H., 'Confucius: His Life and Teaching', *Philosophy*, vol. 26, no. 96, 1951
Hourly History, *Confucius: A Life from Beginning to End*, History of China (independently published, 2017)
Hunter, Michael, *Confucius Beyond the Analects* (Brill, 2017)
Jordan, D. K., 'Confucius: A Brief Biography for College Students', UCSD, September 2019, https://pages.ucsd.edu/~dkjordan/chin/Koong/ConfuciusBio.html
Knechtges, David R., and Chang, Taiping (eds), *Ancient and Early Medieval Chinese Literature: Part 1* (Brill, 2010)
Livaccari, Chris, 'Not Your Father's Confucius', The Asia Society, October 2011, https://asiasociety.org/blog/asia/not-your-fathers-confucius
Moll-Murata, Christine, *State and Crafts in the Qing Dynasty (1644–1911)* (Amsterdam University Press, 2018)
National Geographic Society, 'Confucianism', National Geographic Resource Library, May 2022, https://education.nationalgeographic.org/resource/confucianism
Osnos, Evan, 'Confucius Comes Home', *The New Yorker*, 5 January 2014
Sima Qian, *Records of the Grand Historian*, early 1st century BCE (reproduced by Columbia University Press, 2011)
Stumpfeldt, Hans, 'Thinking Beyond the "Sayings": Comments About Sources Concerning the Life and Teachings of Confucius (551–479)', *Oriens Extremus*, vol. 49, 2010
West, Stephen, 'Confucianism', *Philosophize This!* (podcast), October 2013

3. Leonardo da Vinci

'Leonardo da Vinci', *The Mark Steel Lectures*, BBC TV, 2004
'Leonardo's Harpsichord-Viola', press release, Leonardo3 Museum, Milan, March 2010, https://www.leonardo3.net/data/press_42993/fiche/255/l3_press_release_claviviola_lo-eng_48489.pdf
Durden Smith, Jo, 'Cesare Borgia', *100 Most Infamous Criminals: Murder, Mayhem and Madness* (Arcturus Publishing, 2013)
Figes, Lydia, 'Leonardo da Vinci: a celebration of genius', *Art UK*, 2 May 2019
Finocchi, Nadia, *Leonardo Da Vinci*, English edition (Good Mood, 2015)
Gayford, Martin, 'Was Michelangelo a better artist than Leonardo da Vinci?', *Telegraph*, 16 November 2013
Giaimo, Cara, 'Leonardo da Vinci Designed a Nightmare Scuba Suit', *Atlas Obscura*, July 2016
Hourly History, *Leonardo da Vinci: A Life from Beginning to End*, Biographies of Painters (independently published, 2016)

Isbouts, Jean-Pierre, 'The "Earlier Version" of the Mona Lisa as the Portrait of Lisa del Giocondo described by Vasari', The Mona Lisa Foundation, https://monalisa.org/2013/10/26/the-earlier-version-of-the-mona-lisa-as-the-portrait-of-lisa-del-giocondo-described-by-vasari/

Jones, Daniel J., 'Leonardo da Vinci: Pioneer Geologist', *Brigham Young University Studies*, vol. 4, no. 2, 1962

Keith, Larry, Roy, Ashok, Morrison, Rachel, and Schade, Peter, 'Leonardo da Vinci's *Virgin of the Rocks*: Treatment, Technique and Display', *National Gallery Technical Bulletin, Vol. 32: Leonardo da Vinci: Pupil, Painter and Master*, 2011

Kemp, Martin, 'Leonardo & Michelangelo: rivalry and inspiration', *History Extra*, 13 April 2021

Leonardo da Vinci (trans. Richter, Jean Paul), *The Notebooks of Leonardo da Vinci*, accessed via Wikisource

Lillie, Amanda, 'The lost altarpiece', National Gallery, 2019, www.nationalgallery.org.uk/exhibitions/past/leonardo-experience-a-masterpiece/the-lost-altarpiece

The National Gallery, 'The Virgin of the Rocks', www.nationalgallery.org.uk/paintings/leonardo-da-vinci-the-virgin-of-the-rocks

Vasari, Giorgio (trans. De Vere, Gaston du C.), 'Leonardo Da Vinci (1452–1519)', *Lives of the Most Excellent Painters, Sculptors and Architects* (Macmillan/the Medici Society, 1915)

Vezzosi, Alessandro, and Sabato, Agnese (trans. Frost, Catherine), 'Palazzo Vecchio (formerly Palazzo della Signoria)', Museo Galileo – Istituto e Museo di Storia della Scienza, www.latoscanadileonardo.it/en/places/metropolitan-city-of-florence/municipality-of-florence/palazzo-vecchio.html

4. Galileo

'Galileo Galilei (1564–1642)', *British Journal of Sports Medicine*, vol. 40, no. 9, September 2006

Bendiner, Elmer, 'Renaissance Medicine: Alchemy and Astrology, Art and Anatomy', *Hospital Practice*, vol. 24, no. 6, 1989

Bethune, John Elliot Drinkwater, *The Life of Galileo Galilei with Illustrations of the Advancement of Experimental Philosophy* (1830)

Biagioli, Mario, *Galileo, Courtier: The Practice of Science in the Culture of Absolutism* (University of Chicago Press, 1994)

Bultheel, Adhemar, 'The secret formula', European Mathematical Society, June 2020, https://euro-math-soc.eu/review/secret-formula

Covington, Mark, and Mistry, Amit, 'The Status of Women in Galileo's Time', The Galileo Project, 1995, http://galileo.rice.edu/fam/status_women.html

Danielson, Dennis, and Graney, Christopher M., 'The Case Against Copernicus', *Scientific American*, January 2014

Finocchiaro, Maurice A., '400 Years Ago the Catholic Church Prohibited Copernicanism', *Origins*, Oregon State University, February 2016

Fisher, Len, 'Galileo, Dante Alighieri, and how to calculate the dimensions of hell', ABC News, February 2016

Fisher, Len, *Crashes, Crises, and Calamities: How We Can Use Science to Read the Early-Warning Signs* (Basic Books, 2011)

Inglis-Arkell, Esther, 'Did Galileo get in trouble for being right, or for being a jerk about it?', *Gizmodo*, September 2011

Linton, C. M., *From Eudoxus to Einstein: A History of Mathematical Astronomy* (Cambridge University Press, 2004)

Livio, Mario, *Galileo and the Science Deniers* (Simon & Schuster, 2021)

Magnaghi-Delfino, Paola, and Norando, Tullia, 'Galileo Galilei's location, shape and size of Dante's Inferno: An artistic and educational project', Politecnico di Milano, 2016

Matson, John, 'Galileo's Contradiction: The Astronomer Who Riled the Inquisition Fathered 2 Nuns, a Q&A with author Dava Sobel', *Scientific American*, October 2009

Mottana, Annibale, 'Galileo's La bilancetta: The First Draft and Later Additions', *Philosophia Scientiae* vol. 21, issue 21–1, February 2017

O'Connor, J. J. and Robertson, E. F., 'Galileo Galilei' and 'Nicolaus Copernicus', MacTutor History of Mathematics Archive, University of St Andrews, 2002

Peterson, Mark A., 'Galileo's discovery of scaling laws', *American Journal of Physics*, vol. 70, issue 6, 2002

Radcliff-Umstead, Douglas (ed.), *University World: A Synoptic View of Higher Education in the Middle Ages and Renaissance* (University of Pittsburgh, 1973)

Radford, Tim, 'How Dante beat Galileo to law of motion by 300 years', *Guardian*, 7 April 2005

Santillana, Giorgio, *The Crime of Galileo* (University of Chicago Press, 1955)

Stedall, Jacqueline, 'Medieval and early modern mathematics', University of Cambridge, www.hps.cam.ac.uk/students/research-guide/medieval-early-modern-mathematics

Williams, Matt, 'What is the heliocentric model of the universe?', *Universe Today*, January 2016

5. Tycho Brahe

'A Tale of Fate: From Astrology to Astronomy', *All Things Considered*, National Public Radio, 10 November 2012

'All is not what it seems: the blurred boundaries between alchemy and medicine', University of Cambridge, September 2011, www.cam.ac.uk/research/news/all-is-not-what-it-seems-the-blurred-boundaries-between-alchemy-and-medicine

Charles River Editors, *Tycho Brahe: The Life and Legacy of the Legendary Astronomer Who Mentored Johannes Kepler* (Charles River Editors, 2018)

Ferguson, Kitty, *The Nobleman and His Housedog: Tycho Brahe and Johannes Kepler – The Strange Partnership that Revolutionised Science* (Headline Review, 2002)

Fowler, Michael, 'Tycho Brahe', University of Virginia Physics Department, 1995, https://galileoandeinstein.phys.virginia.edu/1995/lectures/tychob.html

Griggs, Mary Beth, 'We've been predicting eclipses for over 2000 years. Here's how.', *Popular Science*, August 2017

Grossman, Lisa, 'What do plants and animals do during an eclipse?', *Science News*, August 2017

Hernandez, Daisy, 'So Two Moose Are Fighting in Your Driveway Again', *Popular Mechanics*, December 2019

Khan, Razib, 'Inbreeding and the Downfall of the Spanish Hapsburgs', *Discover Magazine*, April 2009

Mancini, Mark, 'Tycho Brahe: The Astronomer with a Drunken Moose', *Mental Floss*, May 2013

Matson, John, 'Was Tycho Brahe poisoned? 16th-century astronomer exhumed – again', *Scientific American*, November 2010

Merrell, Charles, 'Tycho Brahe, The Most Eccentric, Bizarre Man You've Never Heard Of', Academia.edu, 2020

Phillips, Tony, 'What in the World is Hebesphenomegacorona?', NASA, June 2003, www.nasa.gov/audience/forstudents/postsecondary/multimedia/Hebes_mm_feature.html

Rasmussen, K. L., et al., 'Was He Murdered or Was He Not? – Part II: Multi-Elemental Analyses of Hair and Bone Samples from Tycho Brahe and Histopathology of His Bones', *Archaeometry*, vol. 59, issue 5, October 2017

Thoren, Victor E., *The Lord of Uraniborg: A Biography of Tycho Brahe* (Cambridge University Press, 1991)

Tierney, John, 'Murder! Intrigue! Astronomers?', *The New York Times*, 29 November 2010

West, Mary Lou, 'On Tycho's Island: Tycho Brahe and His Assistants, 1570–1601', *Physics Today* 54, 8, 47 (2001)

Wilkins, Alasdair, 'The crazy life and crazier death of Tycho Brahe, history's strangest astronomer', *Gizmodo*, November 2010

6. René Descartes

'René Descartes', *The Mark Steel Lectures*, BBC TV, March 2006

Ariew, Roger, and Grene, Marjorie, 'Ideas, in and before Descartes', *Journal of the History of Ideas*, vol. 56, no. 1, January 1995

Augst, Bertrand, 'Descartes's Compendium on Music', *Journal of the History of Ideas*, vol. 26, no. 1, January–March 1965

Baillet, Adrien, *The Life of Monsieur Des Cartes, Containing the History of his Philosophy and Works: as Also, the Most Remarkable Things that Befell Him During the Whole Course of His Life* (Simpson, 1693)

Browne, Alice, 'Descartes's Dreams', *Journal of the Warburg and Courtauld Institutes*, vol. 40, 1977

Cottingham, John (ed.), *The Cambridge Companion to Descartes* (Cambridge University Press, 1992)

Damjanovic, Aleksandar, Milovanović, Srđan, and Trajanovic, Nikola N., 'Descartes and His Peculiar Sleep Pattern', *Journal of the History of the Neurosciences* 24(4):1-12, August 2015

Descartes, René, *Discourse on the Method of Rightly Conducting One's Reason and of Seeking Truth in the Sciences* (1637)

Edgerton, Samuel Y., 'Heat and Style: Eighteenth-Century House Warming by Stoves', *Journal of the Society of Architectural Historians*, vol. 20, no. 1, March 1961

Gabbey, Alan, 'Descartes' Three Dreams' and 'Rosicrucian', Lawrence Nolan, (ed.), *The Cambridge Descartes Lexicon* (Cambridge University Press, 2016)

Hatfield, Gary, 'René Descartes', Edward N. Zalta (ed.), *The Stanford Encyclopedia of Philosophy* (2018), https://plato.stanford.edu/entries/descartes/

Idowu Otaiku, Abidemi, 'Did René Descartes Have Exploding Head Syndrome?', *Journal of Clinical Sleep Medicine*, April 2018

Melton, J. Gordon, 'Rosicrucian', *Encyclopaedia Britannica*, 2022, www.britannica.com/topic/Rosicrucians

Reilly, Lucas, '17 Things to Know About René Descartes', *Mental Floss*, July 2018

Sasaki, C., *Descartes's Mathematical Thought* (Springer, 2013)

Shea, W. R., 'Descartes and the Rosicrucian Enlightenment', *Metaphysics and Philosophy of Science in the Seventeenth and Eighteenth Centuries*, The University of Western Ontario Series in Philosophy of Science, vol. 43, 1988

Skirry, Justin, 'René Descartes', *Internet Encyclopedia of Philosophy*, https://iep.utm.edu/rene-descartes/

Townsend, Jon, 'Raised Hearth with Oven Paintings', *Savoring the Past*, April 2013

Vallely, Paul, 'Cross purposes: Who are the Rosicrucians?', *Independent*, 6 August 2009

Watson, Richard, *Cogito, Ergo Sum: The Life of René Descartes* (Godine, 2007)

7. Sir Isaac Newton

'From the Library', *British Journal of Ophthalmology*, vol. 87, no. 1308, 2003, https://bjo.bmj.com/content/87/10/1308.info

'Isaac Newton: Physicist And . . . Crime Fighter?', *Science Friday*, National Public Radio, 5 June 2009

'Isaac Newton's Personal Life', The Newton Project, www.newtonproject.ox.ac.uk/his-personal-life

'March 16, 1699: William Chaloner, Counterfeiter, Hanged', *APS Physics*, vol. 20, no. 3, March 2011

Breen, Benjamin, 'Newton's Needle: On Scientific Self-Experimentation', *Pacific Standard*, 24 July 2014

Ducheyne, Steffen, *'The Main Business of Natural Philosophy': Isaac Newton's Natural-Philosophical Methodology* (Springer, 2012)

Fara, Patricia, 'When Isaac Newton was Master of the Royal Mint', *Prospect Magazine*, 21 April 2021

Garisto, Robert, 'An error in Isaac Newton's determination of planetary properties', *American Journal of Physics* vol. 59, no. 42, 1991

Gershon, Livia, 'Isaac Newton Thought the Great Pyramid Held the Key to the Apocalypse', *Smithsonian Magazine*, 8 December 2020

Greenberg, John L., 'Isaac Newton and the Problem of the Earth's Shape', *Archive for History of Exact Sciences*, vol. 49, no. 4, December 1995

Hall, Alfred Rupert, *Philosophers at War: The Quarrel Between Newton and Leibniz* (Cambridge University Press, 2002)

Izadi, Elahe, 'Isaac Newton spent a lot of time on junk "science," and this manuscript proves it', *Washington Post*, 8 April 2016

Lewis, Danny, 'Isaac Newton Used This Recipe in His Hunt to Make a Philosopher's Stone', *Smithsonian Magazine*, 11 April 2016

Litke, James, 'Gravity Still Pulls, Planets Orbit, but Student Catches Newton in Mistake', AP News, 10 June 1987

Maddaluno, Lavinia, 'Four Unpublished Letters from Nicolas Fatio de Duillier to Isaac Newton', Nuncius / Istituto e museo di storia della scienza 34, August 2020

Mann, Adam, 'The Strange, Secret History of Isaac Newton's Papers', *Wired*, 14 May 2014

Matthewson-Grand, Alisha, 'Life under lockdown', University of Cambridge, www.cam.ac.uk/alumni/life-in-lockdown

Mirsky, Steve, 'Inside Isaac: A Discussion of Newton', *Scientific American*, 24 February 2013

Newton, Isaac, 'Of Colours', via The Newton Project, Cambridge University Library, www.newtonproject.ox.ac.uk/view/texts/normalized/NATP00004

Newton, Isaac, 'Quæstiones quædam Philosophiæ' ('Certain Philosophical Questions'), via The Newton Project, Cambridge University Library, www.newtonproject.ox.ac.uk/view/texts/diplomatic/THEM00092

Newton, Isaac, 'Several Questions Concerning the Ph[ilosoph]ers St[one]', from "Keynes MS. 44", *The Chymistry of Isaac Newton*, ed. 2011, http://purl.dlib.indiana.edu/iudl/newton/ALCH00033

Newton, Isaac, 'The Lawes of Motion', via The Newton Project, Cambridge University Library

Philosophical Transactions of the Royal Society, vol. 20, issue 236, 31 January 1698

Samuels, Sean, 'The Story Behind Opticks by Sir Isaac Newton', Bauman Rare Books blog, 9 September 2013

Sastry, S. Subramanya, 'The Newton-Leibniz controversy over the invention of the calculus', University of Wisconsin–Madison, 2004

Swerdlow, N. M., 'Newton's Demonstration of the Force of Gravity and the Definitive Proof of the Heliocentric Theory', *Proceedings of the American Philosophical Society*, vol. 148, no. 1, 2004

8. Wolfgang Amadeus Mozart

'Leopold Mozart, Father of a Prodigy', National Public Radio, 13 January 2006

Ashoori, Aidin, and Jankovic, Joseph, 'Mozart's movements and behaviour: a case of Tourette's syndrome?', *Journal of Neurology, Neurosurgery & Psychiatry*, vol. 78, no. 11, November 2007

Bain, R., et al., 'Fecal Contamination of Drinking-Water in Low- and Middle-Income Countries: A Systematic Review and Meta-Analysis', *PloS Medicine*, vol. 11, no. 5, May 2014

Burton-Hill, Clemency, 'What *Amadeus* gets wrong', *BBC Culture*, 24 February 2015

Clifford, Naomi, 'Breast is best but what were the alternatives in the long 18th century?', 26 September 2017, www.naomiclifford.com/breastfeeding/

Fallon, Brian, 'Mozart's battle with his father', *The Irish Times*, 9 March 1996

Kenyon, Nicholas, *The Faber Pocket Guide to Mozart* (Faber & Faber, 2011)

Keynes, Milo, 'The Personality and Illnesses of Wolfgang Amadeus Mozart', *Journal of Medical Biography*, vol. 2, no. 4, November 1994

McNeill, Fiona E., 'Dying for makeup: Lead cosmetics poisoned 18th-century European socialites in search of whiter skin', *The Conversation*, 27 February 2022

Quirke, Antonia, 'Rude awakening: how Mozart's filthy mind shocked Maggie', *The New Statesman*, 27 November 2014

Reid, Molly, 'Mozart, his mother's death and how it shaped the darkest of his piano sonatas', *The Times-Picayune*, 26 February 2010

Ross, Alex, 'The Storm of Style: Listening to the complete Mozart', *The New Yorker*, 16 July 2006

Suchet, John, *Mozart: The Man Revealed* (Elliott & Thompson Ltd, 2021)

von Tunzelmann, Alex, 'Amadeus: the fart jokes can't conceal how laughably wrong this is', *Guardian*, 22 October 2009

Wallace, Arminta, 'Unravelling the truth about Mozart', *The Irish Times*, 15 June 2004

Wilson, Ransom, 'Mozart: The Boy Genius', Redlands Symphony, 14 January 2017, www.redlandssymphony.com/articles/mozart-the-boy-genius

Wright-Mendoza, Jessie, 'When Breastfeeding Was a Civic Duty', *JSTOR Daily*, 1 October 2018

9. Benjamin Franklin

'December 23, 1750: Ben Franklin Attempts to Electrocute a Turkey', *APS Physics*, vol. 15, no. 11, December 2006

'From Benjamin Franklin to Peter Collinson', 14 August 1747 and 29 April 1749, Founders Online, National Archives

Allain, Rhett, 'Let's Geek Out on the Physics of Leyden Jars', *Wired*, 24 January 2017

Bertucci, Paola, 'Sparks in the dark: the attraction of electricity in the eighteenth century', *Endeavour*, vol. 31, no. 3, September 2007

Gupton, Nancy, 'Benjamin Franklin and the Kite Experiment', The Franklin Institute, 12 June 2017

Inglis-Arkell, Esther, 'The Flying Boy Experiment Entertained Audiences By Electrifying a Kid', *Gizmodo*, 15 January 2015

Marsh, Allison, 'Ben Franklin's Other Great Electrical Discovery: Turkey Tenderization', *IEEE Spectrum*, 30 November 2018
Ouellette, Jennifer, 'That time Benjamin Franklin tried (and failed) to electrocute a turkey', *Ars Technica*, 28 November 2019
Priestley, Joseph, *The History and Present State of Electricity, with Original Experiments* (1767), Founders Online, National Archives

10. Émilie du Châtelet

'December 1706: Birth of Émilie du Châtelet', *APS Physics*, vol. 17, no. 11, December 2008
'Voltaire's Story Revealed in "Passionate Minds"', *Morning Edition*, National Public Radio, 27 November 2006
Allen, Brooke, 'The Multi-Tasking Marquise', *The Hudson Review*, vol. 60, no. 2, Summer 2007
Bodanis, David, 'Honour the woman who enlightened Europe', *Financial Times*, 19 December 2006
Bodanis, David, 'The scientist that history forgot', *Guardian*, 15 May 2006
DeBakcsy, Dale, 'Sex, Cards, and Calculus: A Day with Badass Mathematician Émilie du Châtelet (1706-1749)', *Women You Should Know*, 21 December 2016
Detlefsen, Karen, 'Émilie du Châtelet', Edward N. Zalta (ed.), *The Stanford Encyclopedia of Philosophy* (2018), https://plato.stanford.edu/entries/emilie-du-chatelet/
Eschner, Kat, 'Five Things to Know About French Enlightenment Genius Émilie du Châtelet', *Smithsonian Magazine*, 15 December 2017
Fara, Patricia, 'Emilie du Châtelet: the genius without a beard', *Physics World*, 10 June 2004
Frankenburg, Frances, 'Voltaire and gambling', *International Gambling Studies*, vol. 22, no. 1, 2022
Huffman, Cynthia J., 'Mathematical Treasure: Émilie du Châtelet's Principes Mathématiques', Mathematical Association of America, https://www.maa.org/book/export/html/841298
Janda, Setareh, 'Royal French Manners Were So Weird That You Could Pee Directly in Front of the Queen', *Ranker*, 1 April 2020
Mandic, Sasha, 'Biographies of Women Mathematicians: Emilie du Châtelet', Agnes Scott College, April 1995
Musielak, Dora E., 'The Marquise du Châtelet: A Controversial Woman of Science', University of Texas at Arlington, June 2014
O'Connor, J. J., and Robertson, E. F., 'Gabrielle Émilie Le Tonnelier de Breteuil Marquise du Châtelet', MacTutor History of Mathematics Archive, University of St Andrews, 2003
Strick, Heinz Klaus, 'Émilie du Chatelet (17 December 1706 – 10 September 1749)', trans. O'Connor, John, MacTutor History of Mathematics Archive, University of St Andrews, 2020

Von Baeyer, Hans Christian, review of Bodanis, David, *Passionate Minds: The Great Love Affair of the Enlightenment* (Crown, 2006), *American Journal of Physics*, vol. 75, issue 6, February 2007

Waithe, M. E. (ed.), 'Gabrielle Émilie le Tonnelier de Breteuil du Châtelet-Lomont', *A History of Women Philosophers*, vol 3. (Springer, 1991)

Zinsser, Judith P., *Emilie Du Chatelet: Daring Genius of the Enlightenment* (Penguin Publishing Group, 2007)

11. Johann Christian Reil

'How Victorian Women Were Oppressed Through the Use of Psychiatry', *The Atlantic*, 2017

'Mental illness in the 16th and 17th centuries', Historic England, https://historicengland.org.uk/research/inclusive-heritage/disability-history/1485-1660/mental-illness-in-the-16th-and-17th-centuries/

Bahşi, İ., Adanir, S. S., and Karatepe, Ş., 'Johann Christian Reil (1759–1813) who first described the insula', *Child's Nervous System*, vol. 38, May 2022

Barcella, Laura, 'How Carrie Fisher Championed Mental Health', *Rolling Stone*, 28 December 2016

Beck, Julie, 'Diagnosing Mental Illness in Ancient Greece and Rome', *The Atlantic*, 23 January 2014

Berman, Michele R., 'A Tale of Two Carries: Bipolar Disease and Electroconvulsive Therapy', *Med Page Today*, 1 March 2013

Brunner, Bernd, *Moon: A Brief History* (Yale University, 2010)

Cahalan, Susannah, 'The Reporter Who Went Undercover at an Asylum', *Literary Hub*, 7 November 2019

Dimitrijevic, Aleksandar, 'Being Mad in Early Modern England', *Frontiers in Psychology*, vol. 6, no. 1740, November 2015

Durand, Barlow and Hofmann, *Essentials of Abnormal Psychology* (Wadsworth, 2015)

Goffman, Erving, *Asylums: Essays on the Social Situation of Mental Patients and Other Inmates* (Taylor & Francis, 2017)

Graham, Thomas F., *Medieval Minds: Mental Health in the Middle Ages* (Taylor & Francis, 2019)

Groopman, Jerome, 'The Troubled History of Psychiatry', *The New Yorker*, 20 May 2019

Hansen, LeeAnn, 'Metaphors of Mind and Society: The Origins of German Psychiatry in the Revolutionary Era', *Isis*, vol. 89, no. 3, September 1998

Harrington, Anne, 'A tale of two disorders: syphilis, hysteria and the struggle to treat mental illness', *Nature*, 19 August 2019

Kaplan, Robert M., 'Johann Christian Reil and the naming of our specialty', *Australasian Psychiatry*, vol. 20, no. 2, March 2012

Kirsch, Adam, 'Design for Living', *The New Yorker*, 24 January 2016

Lilienfeld, Scott O., and Arkowitz, Hal, 'The Truth about Shock Therapy', *Scientific American*, 1 May 2014

MacDonald, Scott, and Kretzmann, Norman, 'Medieval philosophy', *Routledge Encyclopedia of Philosophy* (Taylor & Francis, 1998)

Malcolm, Lynne, and Blumer, Clare, 'Madness and insanity: A history of mental illness from evil spirits to modern medicine', ABC News, 2 August 2016

Marneros, A., 'Psychiatry's 200th birthday', *British Journal of Psychiatry*, vol. 193, issue 1, July 2008

McGarvey, Kathleen, 'Goethe was really an outlier in stressing that love was more important', University of Rochester News Center, 18 September 2018

McNamee, David, 'Hey, what's that sound: the Katzenklavier', *Guardian*, 19 April 2010

Munson-Barkshire, Amy, 'Scandals in Chronic Sector Hospitals' (dissertation), Socialist Health Association, March 1981

Reil, Johann C., 'kleine Schriften wissenschaftlichen und gemeinnützigen Inhalts: Mit 1 Kupf', via https://www.digitale-sammlungen.de/en/view/bsb10085926?page=10,11

Reil, Johann Christian, 'Rhapsodieen über die Anwendung der psychischen Curmethode auf Geisteszerrüttungen', 1803

Richards, Robert J., 'Rhapsodies on a Cat-Piano, or Johann Christian Reil and the Foundations of Romantic Psychiatry', *Critical Inquiry*, vol. 24, no.3, Spring 1998

Richards, Robert J., *The Romantic Conception of Life: Science and Philosophy in the Age of Goethe* (University of Chicago Press, 2010)

Rosen, George, 'Social Attitudes to Irrationality and Madness in 17th and 18th Century Europe', *Journal of the History of Medicine and Allied Sciences*, vol. 18, issue 3, July 1963

Sadowsky, Jonathan, 'Electroconvulsive therapy: A history of controversy, but also of help', *The Conversation*, 13 January 2017

Schochow, Maximilian, and Steger, Florian, 'Johann Christian Reil (1759–1813): Pioneer of Psychiatry, City Physician, and Advocate of Public Medical Care', *The American Journal of Psychiatry*, vol. 171, issue 4, April 2014

Schui, Florian, *Rebellious Prussians: Urban Political Culture Under Frederick the Great and His Successors* (OUP, 2013)

Vann, Madeline R., and Marcellin, Lindsey, 'The 10 Worst Mental Health Treatments in History', *Everyday Health*, 7 May 2014

Wallace, Edwin R., and Gach, John (eds), *History of Psychiatry and Medical Psychology* (Springer US, 2008)

12. Napoleon Bonaparte

'Campaigns and Battles: Napoleon at War', Public Broadcasting Service, www.pbs.org/empires/napoleon/n_war/campaign/page_8.html

'Music In Time: Napoleon vs. Rabbits', ABC Classic, broadcast 1 April 2019

'Napoleon and Rabbits', *Liverpool Herald* (New South Wales), 6 April 1901, http://nla.gov.au/nla.news-article37254325

Cavendish, Richard, 'The Treaty of Tilsit', *History Today*, vol. 57, issue 7, July 2007

Edwardes, Charlotte, 'Historian obsessed with Napoleon spills the beans on

Bonaparte's sex life and reveals the truth about "not tonight, Josephine"', *Evening Standard*, 10 June 2015

Raub, Olivia, 'Napoleon's Greatest Failure (No, Not Waterloo)', History as You were Never Taught, Penn State University, 20 March 2014

13. Lord Byron

'Byron as a Boy', *The New York Times*, 26 February 1898, via TimesMachine

'History's ultimate "crazy ex girlfriend": Caroline Lamb', F Yeah History, June 2016

'Letter from Lord Byron to John Murray about the death of Keats, 26 April 1821', British Library

'Letter to Douglas Kinnaird from George Gordon, Lord Byron, 26 October 1819', http://mason.gmu.edu/~rnanian/Byron-Kinnaird.html

'Lord Byron (George Gordon)', Poetry Foundation

'Lord Byron', *The Mark Steel Lectures*, BBC TV, October 2003

'Review: Letters and Journals of Lord Byron; with Notices of his Life, by Thomas Moore, Anonymous', *The London Literary Gazette and Journal of Belles Lettres, Arts, Sciences &c.*, no. 678, 16 January 1830

Brand, Emily, 'Sex, swords and incest: the many scandals of "Mad Jack" Byron', *Historia Magazine*, 15 February 2021

Brand, Emily, *The Fall of the House of Byron: Scandal and Seduction in Georgian England* (John Murray, 2021)

Brown, Mark, 'The Byronic look: overweight and unattractive', *Guardian*, 26 August 2011

Byron, George Gordon, *The Complete Works of Lord Byron Including His Suppressed Poems and Others Never Before Published*, vol. 4 (Baudry's Foreign Library, 1832)

Byron, George Gordon (ed. Coleridge, Prothero), *The Works of Lord Byron* (John Murray, 1903)

Byron, George Gordon, and Marchand, Leslie Alexis, *Byron's Letters and Journals: The Complete and Unexpurgated Text of All the Letters Available in Manuscript and the Full Printed Version of All Others* (Belknap Press of Harvard University Press, 1973)

Castelow, Ellen, 'Lord Byron', Historic UK, https://www.historic-uk.com/CultureUK/Lord-Byron/

Cochran, Peter (ed.), 'Byron Courts Annabella Milbanke: Correspondences August 1813–December 1814', March 2010, https://petercochran.files.wordpress.com/2010/03/byron-and-annabella-1813-1814.pdf

Cochran, Peter, 'Why Did Byron Hate Southey?', a paper read to the Newstead Abbey Byron Society, http://www.newsteadabbeybyronsociety.org/works/downloads/byron_southy.pdf

Cooper, Glenda, 'Byron was severely anorexic', *Independent*, 8 October 1998

Douglass, Paul, 'The Madness of Writing: Lady Caroline Lamb's Byronic Identity', San Jose State University, January 1999

Drummond, Clara, 'Lord Byron, 19th-century bad boy', British Library, 15 May 2014

Forward, Stephanie, 'Lord Byron's "Love and Gold"', British Library
Galt, John, *The Life of Lord Byron* (Henry Colburn and Richard Bentley, 1830)
Grosskurth, Phyllis, *Byron: The Flawed Angel* (Sceptre, 1997)
Gulliver, Lindsay, and Matthew, Adam, 'Shake not your heads, nor say the Lady's mad: A very Byronic bonfire', Adam Matthew Digital, 6 November 2020
Harvey, A. D., 'Prosecutions for Sodomy in England at the Beginning of the Nineteenth Century', *The Historical Journal*, vol. 21, no. 4, December 1978
Larman, Alexander, 'Byron and his women: Mad, bad and very dangerous to know', Wordsworth.org.uk, 8 September 2016
Macaulay, Susy, 'Byron's formative years in Aberdeen – did they make him mad, bad and dangerous?', *The Press and Journal*, 10 August 2021
MacCarthy, Fiona, 'In bed with Byron', *Guardian*, 30 May 1999
MacCarthy, Fiona, 'Poet of all the passions', *Guardian*, 9 November 2002
MacCarthy, Fiona, *Byron: Life and Legend* (John Murray, 2014)
Maye, Brian, '"Here lie the bones of Castlereagh" – An Irishman's Diary on a political colossus', *Irish Times*, 17 June 2019
McGann, Jerome, 'Byron, George Gordon Noel, sixth Baron Byron', *Oxford Dictionary of National Biography*, 24 October 2019
Medwin, Thomas, *Journal of the Conversations of Lord Byron: Noted During a Residence with His Lordship at Pisa in the Years 1821 and 1822* (Wilder & Campbell, 1824)
Moore, Thomas (ed.), 'Letter from Lord Byron to Francis Hodgson, 3 October 1810', *Letters and Journals of Lord Byron* (John Murray, 1830)
Norton, Rictor (ed.), 'The Gay Love Letters of Lord Byron: Excerpts from *My Dear Boy: Gay Love Letters through the Centuries*', 1998, https://rictornorton.co.uk/byron.htm
Robertson, J. Michael, 'Aristrocratic [sic] Individualism in Byron's Don Juan', *Studies in English Literature, 1500–1900*, vol. 17, no. 4, Autumn 1977
Robertson, Michael, 'The Byron of Don Juan as Whig Aristocrat', *Texas Studies in Literature and Language*, vol. 17, no. 4, Winter 1976
Rutigliano, Olivia, 'Lord Byron used to call William Wordsworth "Turdsworth," and yes, this is a real historical fact', *Literary Hub*, 14 January 2020
Schiff, Stacy, 'Mad, Bad and Dangerous: The Legacy Left to Byron's Wife and Daughter', *The New York Times*, 30 November 2018
Stark, Myra, 'The Princess of Parallelograms, or the Case of Lady Byron', *Keats-Shelley Journal*, vol. 31, 1982
Tonkin, Boyd, 'Mad, bad and delightful to know: How Lord Byron became a cultural superstar', *Independent*, 16 February 2012
Wilson, Cheryl A. (ed.), *Byron: Heritage and Legacy* (Palgrave Macmillan, 2008)

14. Ada Lovelace

Charman-Anderson, Suw (ed.), *A Passion for Science: Tales of Discovery and Invention* (FindingAda, 2015)

Fry, Hannah, 'Not your typical role model: Ada Lovelace the 19th century programmer', BBC, 2015

Hourly History, *Ada Lovelace: A Life from Beginning to End*, Biographies of Women in History (independently published, 2019)

Klein, Christopher, '10 Things You May Not Know About Ada Lovelace', History.com, 22 August 2018

Kleinman, Zoe, 'Ada Lovelace: Opium, Maths and the Victorian programmer', BBC News, 12 October 2015

Monahan, John, *Babbage & Lovelace: The Victorian Computer Wizard and the Enchantress of Numbers* (independently published, 2019)

Morais, Betsy, 'Ada Lovelace, the First Tech Visionary', *The New Yorker*, 15 October 2013

Stoffer, Shawn, 'Ada Lovelace', University of New Mexico, Fall 1999

Toole, Betty Alexandra, 'A Selection and Adaptation from Ada's Notes found in *Ada, The Enchantress of Numbers*' (Strawberry Press, 1998), https://www.cs.yale.edu/homes/tap/Files/ada-lovelace-notes.html

Wolfram, Stephen, 'Untangling the Tale of Ada Lovelace', *Wired*, 22 December 2015

Woolley, Benjamin, *The Bride of Science: Romance, Reason and Byron's Daughter* (Pan, 2015)

15. Évariste Galois

'19th century: thrust into the upheaval of the times', Institut Polytechnique de Paris, www.polytechnique.edu/en/school/history/19th-century-thrust-upheaval-times

'Mathematician Jordan Ellenberg breaks down math films & TV shows', Penguin Books UK, www.youtube.com/watch?v=oejBbQx-Sqo

Ayel, Mathieu, 'The French Grandes Écoles: The Revolution and the École Polytechnique', MacTutor History of Mathematics Archive, University of St Andrews

du Sautoy, Marcus, 'A Brief History of Mathematics: Évariste Galois', BBC Radio 4, 17 June 2010

Galois, Évariste, 'Preface', December 1831, via https://mathshistory.st-andrews.ac.uk/Extras/Galois_Sainte_Pelagie_preface/

Kruglinski, Susan, 'Mysterious Death of a Mathematician Finally Solved?', *Discover Magazine*, 21 November 2005

Lienhard, John H., 'The Engines of Our Ingenuity: Évariste Galois', University of Houston, 1988-99, https://www.uh.edu/engines/epi1475.htm

Livio, Mario, *The Equation that Couldn't Be Solved: How Mathematical Genius Discovered the Language of Symmetry* (Simon & Schuster, 2005)

Lynch, Peter, 'The tragic, brief life of Évariste Galois', *The Irish Times*, 15 August 2019

Miller, Andrew, 'My hero: Évariste Galois', *Guardian*, 20 January 2012

O'Connor, J. J., and Robertson, E. F., 'Évariste Galois', MacTutor History of Mathematics Archive, University of St Andrews

Omitola, Tope, 'Genius, stupidity and genius again', *Plus*, 1 May 2005
Rothman, Tony, 'Genius and Biographers: The Fictionalization of Évariste Galois', *Mathematical Association of America*, vol. 89, no. 2, February 1982

16. John Couch Adams

Airy, George Biddell, 'Account of some circumstances historically connected with the discovery of the planet exterior to Uranus', *Monthly Notices of the Royal Astronomical Society*, vol. 7, November 1846
Bramley, Chris, 'Once round the Sun for Neptune', *The Sky at Night Magazine*, 12 July 2011
Gauss, C. F., and Schumacher, H. C., *Briefwechsel zwischen C. F. Gauss und H. C. Schumacher*, vol. 5 (C. A. F. Peters, 1863)
Kennett, Carolyn, Smith, Robert, Bell, Trudy E., and Sheehan, William (eds), *Neptune: From Grand Discovery to a World Revealed: Essays on the 200th Anniversary of the Birth of John Couch Adams* (Springer International, 2021)
Kollerstrom, Nicholas, 'An Hiatus in History: The British Claim for Neptune's Co-Prediction, 1845–1846: Part 1', *History of Science*, vol. 44, issue 1, March 2006
Kollerstrom, Nicholas, 'John Herschel on the Discovery of Neptune', *Journal of Astronomical History and Heritage*, vol. 9, no. 2, December 2006
Kollerstrom, Nicholas, 'Recovering the Neptune files', *Astronomy & Geophysics*, vol. 44, issue 5, October 2003
Lienhard, John H., 'The Engines of Our Ingenuity: The Neptune Affair', University of Houston, 1988-97, https://www.uh.edu/engines/epi1006.htm
Linton, C. M., *From Eudoxus to Einstein: A History of Mathematical Astronomy* (Cambridge University Press, 2004)
Sheehan, William, Kollerstrom, Nicholas, and Waff, Craig B., 'The Case of the Pilfered Planet', *Scientific American*, vol. 291, no. 6, December 2004
Smith, Robert W., 'The Cambridge Network in Action: The Discovery of Neptune', *Isis*, vol. 80, no. 3, September 1989
Standage, Tom, *The Neptune File: A Story of Astronomical Rivalry and the Pioneers of Planet Hunting* (Berkley, 2001)

17. Karl Marx

'Karl Marx', *The Mark Steel Lectures*, BBC TV, November 2003
Fluss, Harrison, and Miller, Sam, 'The Life of Jenny Marx', *Jacobin*, February 2016
Heinrich, Michael, *Karl Marx and the Birth of Modern Society: The Life of Marx and the Development of His Work*, vol. 1 (Monthly Review Press, 2019)
Holmes, Rachel, 'Karl Marx: the drinking years', *Telegraph*, 14 October 2017
Hunt, Tristram, *The Frock-Coated Communist: The Revolutionary Life of Friedrich Engels* (Penguin, 2009)
Liebknecht, Wilhelm, *Karl Marx: Biographical Memoirs* (First German edition, Nuremberg, 1896; first English translation, by E. Untermann, 1901. Reprinted by Journeyman Press, London, 1975)

Marx, Karl, 'Confessions', April 1865, first published in the *International Review of Social History*, 1956, via www.marxists.org/archive/marx/works/1865/04/01.htm

Wheen, Francis, *Karl Marx: A Life* (W. W. Norton & Company, 1999)

18. Charles Darwin

'Charles Darwin', *The Mark Steel Lectures*, BBC TV, November 2003

'Darwin's Very Bad Day: "Oops, We Just Ate It!"', *All Things Considered*, National Public Radio, 24 February 2009

'Former Governor FitzRoy commits suicide', New Zealand History (Ministry for Culture and Heritage), October 2021, https://nzhistory.govt.nz/page/former-governor-fitzroy-commits-suicide

Askew, David, 'The Gentleman Naturalist', *Dublin Review of Books*, June 2013

Browne, E. Janet, *Charles Darwin: The Power of Place*, vol. 2 (Princeton University Press, 2003)

Carroll, Kathleen, 'From Not So Simple a Beginning: The Voyage of the *Beagle* to the Voyage of the *Endeavour*', University of Maine, December 2012

Chambers, Paul, *A Sheltered Life: The Unexpected History of the Giant Tortoise* (Oxford University Press, 2004)

Darwin, Charles, *The Autobiography of Charles Darwin 1809–1882* (Collins, 1958)

Darwin, Charles, *The Life and Letters of Charles Darwin, Including an Autobiographical Chapter: Vol. I* (John Murray, 1887)

Deaton, Jeremy, 'The tragic story of the founder of weather forecasting in Victorian England', *Washington Post*, 25 April 2021

Dzombak, Rebecca, 'Tiny worms "hear" without an eardrum, surprising scientists', *National Geographic*, 6 October 2021

Hennessy, Elizabeth, 'Charles Darwin, Tortoise Hunter?', Yale University Press, 18 November 2019

Inglis-Arkell, Esther, 'What Did Charles Darwin Put in His Mouth? Pretty Much Everything', *Gizmodo*, 24 February 2015

Keynes, Richard Darwin (ed.), *Charles Darwin's Beagle Diary* (Cambridge University Press, 2001)

MacDonald, James, 'Bombardier Beetles Are Terrifying Nightmare Insects', *JSTOR Daily*, 25 February 2018

McKie, Robin, 'Man on a suicide mission', *Observer*, 29 June 2003

Moore, Peter, 'The birth of the weather forecast', BBC News, 30 April 2015

Moore, Peter, 'The Tragic Life of Charles Darwin's Captain', *History Today*, vol. 65, issue 6, June 2015

Noyce, Diana, 'Charles Darwin, the Gourmet Traveler', *Gastronomica*, vol. 12, no. 2, Summer 2012

Oppenheimer, Mark, 'Seeing Darwin Through Christian Eyes? It All Depends on the Christian', *The New York Times*, 1 February 2013

Palca, Joe, 'Darwin's Earthworm Experiments Broke New Ground,' *All Things Considered*, National Public Radio, 12 February 2009

Paul, Diane B., Stenhouse, John, and Spencer, Hamish G., 'The Two Faces of Robert FitzRoy, Captain of HMS *Beagle* and Governor of New Zealand', *The Quarterly Review of Biology*, vol. 88, no. 3, September 2013

Rack, Jessie, 'Dining Like Darwin: When Scientists Swallow Their Subjects', *The Salt*, National Public Radio, 12 August 2015

Serena, Katie, 'Charles Darwin Not Only Discovered Species, He Also Ate Them In A Glutton Club', *All That's Interesting*, 30 November 2017

Sulloway, Frank J., 'The Evolution of Charles Darwin', *Smithsonian Magazine*, December 2005

19. James Glaisher

'Another Scientific Balloon Ascent', *The Times*, 6 September 1862, via The Times Archive

'History of Ballooning', National Balloon Museum, www.nationalballoonmuseum.com/about/history-of-ballooning/

'Mr Coxwell's Mammoth Balloon', *The Times*, 23 August 1862, via The Times Archive

'Obituary: Mr James Glaisher', *The Times*, 9 February 1903, via The Times Archive

Coxwell, Henry Tracey, *My Life and Balloon Experiences: With a Supplementary Chapter on Military Ballooning*, vol. 2 (W. H. Allen & Company, 1889)

Hickie, Catherine, 'A Fatal Form of Contentment', *The Permanente Journal*, vol. 13, no. 2, June 2009

Glaisher, James, et al., *Travels in the Air* (R. Bentley & Son, 1871)

Perry, Harry, 'The Gasification of Coal', *Scientific American*, vol. 230, no. 3, March 1974

Robson, David, 'The Victorians who flew as high as jumbo jets', *BBC Future*, 20 April 2016

Smith, Zoe, 'Disaster at 37,000 feet', University of Cambridge, https://www.cam.ac.uk/stories/BalloonDisaster

Tucker, Jennifer, 'From their balloons, the first aeronauts transformed our view of the world', *The Conversation*, 6 December 2019

Tucker, Jennifer, 'Voyages of Discovery on Oceans of Air: Scientific Observation and the Image of Science in an Age of "Balloonacy"', *Osiris*, vol. 11, no. 1, 1996

20. Sigmund Freud

'A Tale Of Two Addicts: Freud, Halsted And Cocaine', *Talk of the Nation*, National Public Radio, 25 November 2011

'Cocaine: How "Miracle Drug" Nearly Destroyed Sigmund Freud [and] William Halsted', *PBS News Hour*, 17 October 2011

'Sigmund Freud', *The Mark Steel Lectures*, BBC TV, October 2003

Bernfeld, Siegfried, 'Freud's Studies on Cocaine, 1884–1887', *Journal of the American Psychoanalytic Association*, vol. 1, issue 4, October 1953

Borch-Jacobsen, Mikkel, 'Ernst Fleischl von Marxow (1846–1891)', *Psychology Today*, 7 February 2012

Cohen, David, *Freud on Coke* (Andrews UK, 2011)

Crews, Frederick, 'How Sigmund Freud Tried to Break and Remake His Fiancée', *Literary Hub*, 22 August 2017

Diski, Jenny, 'The Housekeeper of a World-Shattering Theory', *London Review of Books*, vol. 28, no. 6, March 2006

Freud, Ernst L., *Letters of Sigmund Freud* (Dover, 1960)

Freud, Sigmund, 'Letters From Freud to Jung' (from *Memories, Dreams and Reflections*), 16 April 1909, via https://understandinguncertainty.org/user-submitted-coincidences/freud-was-haunted-certain-numbers

Freud, Sigmund, *The Interpretation of Dreams* (Brill, November 1899)

Lennard, Henry L., Freud's Disaster with Cocaine, *The New York Times*, 22 July 1972, via TimesMachine

Markel, Howard, *An Anatomy of Addiction: Sigmund Freud, William Halsted, and the Miracle Drug Cocaine* (Pantheon, 2011)

Mills, James, review of Markel, Howard, *An Anatomy of Addiction* (Vintage, 2012), vol. 27, no. 2, Summer 2013

Masson, Jeffrey Moussaieff, *The Assault on Truth* (Untreed Reads Publishing, 2012)

Oliver, Scott, 'A Brief History of Freud's Love Affair with Cocaine', *Vice*, 23 June 2017

Steiner, Riccardo, 'Die Brautbriefe: the Freud and Martha correspondence', *The International Journal of Psychoanalysis*, vol. 94, issue 5, 2013

Teusch, Rita K., 'More Courtship Letters of Freud and Martha Bernays', *Journal of the American Psychoanalytic Association*, vol. 65, issue 1, February 2017

Trimarchi, Matteo, et al., 'The disease of Sigmund Freud: oral cancer or cocaine-induced lesion?', *European Archives of Oto-Rhino-Laryngology*, vol. 276, issue 1, January 2019

21. Arthur Conan Doyle

'Cottingley Fairies: How Sherlock Holmes's creator was fooled by hoax', BBC News, 5 December 2020

'Fooled by fairies', University of Leeds Library News, 7 December 2020

'How Sir Arthur Conan Doyle's student days at Edinburgh University brought Sherlock Holmes to life', BBC Scotland, 22 May 2017

'The Most Important Thing in the World', https://arthurconandoyle.co.uk/spiritualist

Chalmers, John, 'Conan Doyle And Joseph Bell: The Real Sherlock Holmes', *Independent*, 7 August 2006

Clark Godbee, Dan, 'Joseph Bell (1837–1911): A Clinician's Literary Legacy', *Journal of Medical Biography*, vol. 7, issue 3, August 1999

Conan Doyle, Arthur, *Adventures of Sherlock Holmes* (October 1892)

Conan Doyle, Arthur, 'Fairies Photographed: An Epoch-Making Event', *The Strand Magazine*, December 1920

Conan Doyle, Arthur, *The Stark Munro Letters* (Longmans, Green & Co, 1895)
Conan Doyle, Arthur, *The Wanderings of a Spiritualist* (Hodder & Stoughton, 1921)
Gardner, Lyn, 'Harry Houdini and Arthur Conan Doyle: a friendship split by spiritualism', *Guardian*, 10 August 2015
Gosden, Chris, 'Why Harry Houdini DID NOT Like Arthur Conan Doyle: When the Rational Trickster Meets the Credulous Rationalist', *Literary Hub*, 19 November 2020
Lellenberg, Jon, Stashower, Daniel, and Foley, Charles, *Arthur Conan Doyle: A Life in Letters* (Penguin, September 2007)
Lyster, Rosa, 'The Cottingley fairy hoax of 1917 is a case study in how smart people lose control of the truth', *Quartz*, 17 February 2017
Miller, Russell, *The Adventures of Arthur Conan Doyle* (Random House, June 2010)
Sandford, Christopher, 'When Sir Arthur Conan Doyle (maybe) spoke with the dead', *America Magazine*, 19 November 2020
Snailham, Fiona, and Barry, Anna Maria, 'The supernatural interests of Sir Arthur Conan Doyle', *History Extra*, 8 February 2022
Tompkins, Matthew, 'The two illusions that tricked Arthur Conan Doyle', *BBC Future*, 30 August 2019

22. Thomas Edison

'Business: The Quintessential Innovator', *TIME*, 22 October 1979
'Can You Pass the Test Thomas Edison Gave to His Potential Employees?', *Interesting Engineering*, 12 April 2017
'The Thomas A. Edison Papers Project', Rutgers School of Arts and Sciences, https://edison.rutgers.edu/
'Thomas Edison's "lost" idea: A device to hear the dead', AFP, 5 March 2015, https://phys.org/news/2015-03-thomas-edison-lost-idea-device.html
US Patent US223898A, Inventor: Thomas Alva Edison, via https://patents.google.com/patent/US223898
Cep, Casey, 'The Real Nature of Thomas Edison's Genius', *The New Yorker*, 21 October 2019
Daugherty, Greg, 'Talking to the Dead: How the 1918 Pandemic Spurred a Spiritualism Craze', History.com, 21 April 2020
Edwards, Phil, 'Tesla vs. Edison – and what the never-ending battle says about us', *Vox*, 21 July 2015
Galant, Debra, 'The Father of Invention', *The New York Times*, 1 June 1997
Giaimo, Cara, 'Making Fun of Thomas Edison', *Atlas Obscura*, 28 February 2017
Hendry, Erica R., '7 Epic Fails Brought to You By the Genius Mind of Thomas Edison', *Smithsonian Magazine*, 20 November 2013
Knapp, Alex, 'Nikola Tesla Wasn't God And Thomas Edison Wasn't The Devil', *Forbes*, 18 May 2012
Latson, Jennifer, 'How Edison Invented the Light Bulb – And Lots of Myths About Himself', *TIME*, 21 October 2014

McAuliffe, Kathleen, 'The Undiscovered World of Thomas Edison', *The Atlantic*, December 1995

Morris, Edmund, *Edison* (Random House, 2019)

Morus, Iwan, 'Thomas Edison: visionary, genius or fraud?', *The Conversation*, 12 July 2018

Nash, David, 'The rise of spiritualism after WW1', *History Extra*, 3 May 2020

Palermo, Elizabeth, and McKelvie, Callum, 'Who invented the lightbulb?', *Live Science*, 23 November 2021

Stross, Randall, *The Wizard of Menlo Park: How Thomas Alva Edison Invented the Modern World* (Crown Publishers, 2007)

Tablang, Kristin, 'Thomas Edison, B.C. Forbes and the Mystery of the Spirit Phone', *Forbes*, 25 October 2019

Thompson, Derek, 'Thomas Edison's Greatest Invention', *The Atlantic*, November 2019

Waxman, Olivia B., 'The Real History Behind The Current War', *TIME*, 25 October 2019

Wilkes, Jonny, 'Edison, Westinghouse and Tesla: the history behind The Current War', *History Extra*, March 2019

Zarrelli, Natalie, 'Dial-a-Ghost on Thomas Edison's Least Successful Invention: the Spirit Phone', *Atlas Obscura*, 18 October 2016

23. Nicola Tesla

'Colorado Springs, Tesla: Life and Legacy', Public Broadcasting Service, https://www.pbs.org/tesla/ll/ll_colspr.html

'Science: Tesla at 75', *TIME*, 20 July 1931

Cheney, Margaret, *Tesla: Man Out of Time* (Simon and Schuster, 2001)

Cheney, Margaret, Uth, Robert, and Glenn, Jim, *Tesla, Master of Lightning* (Barnes and Noble, 1999)

Gunderman, Richard, 'Nikola Tesla: The extraordinary life of a modern Prometheus', *The Conversation*, 3 January 2018

King, Gilbert, 'The Rise and Fall of Nikola Tesla and His Tower', *Smithsonian Magazine*, 4 February 2013

Novak, Matt, 'Nikola Tesla's Amazing Predictions for the 21st Century', *Smithsonian Magazine*, 19 April 2013

Tesla, Nikola, as told to George Sylvester Viereck, 'A Machine to End War: A Famous Inventor, Picturing Life 100 Years from Now, Reveals an Astounding Scientific Venture Which He Believes Will Change the Course of History', *Liberty*, February 1937, via https://www.pbs.org/tesla/res/res_art11.html

Turi, J., 'Tesla's toy boat: A drone before its time', Engadget, 19 January 2014

Vaughan, Don, 'Nikola Tesla's Weird Obsession with Pigeons', *Encyclopaedia Britannica*, 2020, www.britannica.com/story/nikola-teslas-weird-obsession-with-pigeons

24. Marie Curie

'March 1, 1896: Henri Becquerel Discovers Radioactivity', *APS Physics*, vol. 17, no. 3, March 2008

'Medicine: Radium Drinks', *TIME*, 11 April 1932

Clark Estes, Adam, 'Marie Curie's century-old radioactive notebook still requires lead box', *Gizmodo*, 4 August 2014

Davies, Norman, *God's Playground A History of Poland, Volume II: 1795 to the Present* (OUP Oxford, 2005)

Des Jardins, Julie, 'Madame Curie's Passion', *Smithsonian Magazine*, October 2011

Gasińska, Anna, 'Life and Work of Marie Sklodowska-Curie and her Family', *Acta Oncologica*, vol. 38, issue 7, 1999

Parsons, Eleanor, 'Marie Curie', *New Scientist*, undated

Pasachoff, Naomi, 'Marie Curie and the Science of Radioactivity: Polish Girlhood', American Institute of Physics, 2000

Prisco, Jacopo, 'When beauty products were radioactive', CNN, 9 March 2020

Pszenicki, Chris, 'The Flying University', *Index on Censorship*, vol. 8, no. 6, November 1979

25. Albert Einstein

'Albert Einstein', Southold Historical Museum, https://www.southoldhistorical.org/albert-einstein

'Einstein Expounds His New Theory', *The New York Times*, 3 December 1919, via TimesMachine

'Einstein Has Fled Temporarily from Germany Because of Threats that He Will Be Killed', *The New York Times*, 6 August 1922, via TimesMachine

'Einstein Not for the Few', *The New York Times*, 28 August 1921, via TimesMachine

'Einstein Restates Relativity Theory', *The New York Times*, 21 May 1920, via TimesMachine

'Einstein Sees End of Time and Space', *The New York Times*, 4 April 1921, via TimesMachine

'Einstein Theory "Bourgeois" and Dangerous, Say Russians', *The New York Times*, 16 November 1922, via TimesMachine

'Jazz in Scientific World', *The New York Times*, 16 November 1919, via TimesMachine

'Russia: Einstein Friend Not a Spy', AP News, 3 June 1998

Akbar, Arifa, 'Einstein's theory of infidelity', *Independent*, 11 July 2006

Arntzenius, Linda G., *Institute for Advanced Study* (Arcadia, 2011)

Burrow, Gerard N., *A History of Yale's School of Medicine: Passing Torches to Others* (Yale University Press, 2008)

Calaprice, Alice, and Lipscombe, Trevor, *Albert Einstein: A Biography* (Greenwood Press, 2005)

Clark, Ronald, *Einstein: The Life and Times* (Bloomsbury, 2011)

Dreier, Peter, 'Was Albert Einstein a Racist?', *The American Prospect*, 19 June 2018

Falk, Dan, 'One Hundred Years Ago, Einstein Was Given a Hero's Welcome by America's Jews', *Smithsonian Magazine*, 2 April 2021

Fantova, Johanna, trans. Calaprice, Alice, 'From the profound to the mundane – diary provides a slice of life', *Princeton Weekly Bulletin*, vol. 93, no. 25, April 2004

Francis, Matthew, 'How Albert Einstein Used His Fame to Denounce American Racism', *Smithsonian Magazine*, 3 March 2017

Gagnon, Pauline, 'The Forgotten Life of Einstein's First Wife', *Scientific American*, 19 December 2016

Galison, Peter, 'Removing Knowledge: The Logic of Modern Censorship', in Proctor, Robert N., and Schiebinger, Londa (eds), *Agnotology: The Making and Unmaking of Ignorance* (Stanford University Press, 2008)

Gewertz, Ken, 'Albert Einstein, Civil Rights activist', *The Harvard Gazette*, 12 April 2007

Kilgannon, Corey, 'Recalling Albert Einstein, who was not much of a genius when it came to sailing', *The New York Times*, 22 July 2007

Libman, E., and Maloney, Russell, 'Disguise', *The New Yorker*, 6 January 1939

Marek, Tomas, 'Albert Einstein: the passionate sailor', Yachting.com, 29 March 2022

Missner, Marshall, 'Why Einstein Became Famous in America', *Social Studies of Science*, vol. 15, no. 2, May 1985

Palmer, Alex, 'A Century Ago, Einstein's First Trip to the U.S. Ended in a PR Disaster', *Discover Magazine*, 24 April 2021

Payne, Kenneth W., 'Einstein on Americans', 10 July 1921, via TimesMachine

Pogrebin, Robin, 'Love Letters by Einstein at Auction', *The New York Times*, 1 June 1998

Popova, Maria, 'Albert Einstein's Little-Known Correspondence with W. E. B. Du Bois About Equality and Racial Justice', *The Marginalian*, 6 January 2015

Pruitt, Sarah, 'Einstein Had No Clue His Lover Was a Suspected Russian Spy', History.com, 23 June 2017

Schiott, J., 'The Einstein Theory', *The New York Times*, 4 April 1920, via TimesMachine

Schultz, Steven, 'Newly discovered diary chronicles Einstein's last years', *Princeton Weekly Bulletin*, vol. 93, no. 25, April 2004

Shriver, Jean, 'My Turn: Albert Einstein: Just a neighbor to this little girl', *Daily Breeze*, 10 June 2014

Smith, Dinitia, 'Dark Side of Einstein Emerges in His Letters', *The New York Times*, 6 November 1996

Troy, Michael, 'How a love of sailing helped Einstein explain the universe', ABC News, 24 November 2017

26. Kurt Gödel

Aziz, Omer, 'The Incompleteness of Everything: Kurt Gödel and the Mathematics of the Unknown', Notes From the Margins, 13 May 2020, https://omerazizwriter.substack.com

Budiansky, Stephen, *Journey to the Edge of Reason: The Life of Kurt Gödel* (OUP, 2021)
Goldstein, Rebecca, *Incompleteness: The Proof and Paradox of Kurt Gödel* (W. W. Norton & Company, 2006)
Holt, Jim, 'Time Bandits', *The New Yorker*, 20 February 2005
Ings, Simon, 'Waiting for Gödel is over: the reclusive genius emerges from the shadows', *Spectator World*, 27 May 2021
Veisdal, Jørgen, 'Kurt Gödel's Brilliant Madness', *Privatdozent*, 21 June 2021

27. Maya Angelou

'At 80, Maya Angelou Reflects on a "Glorious" Life', *Weekend Edition*, National Public Radio, 6 April 2008
Alexandra, Rae, 'A Century Before Rosa Parks, She Fought Segregated Transit in SF', KQED, 12 March 2019
Angelou, Maya, *I Know Why the Caged Bird Sings* (Virago, 1984)
Angelou, Maya, *Mom & Me & Mom* (Virago, 2013)
Fox, Margalit, 'Maya Angelou, Lyrical Witness of the Jim Crow South, Dies at 86', *The New York Times*, 28 May 2014
Nichols, John, 'Maya Angelou's Civil Rights Legacy', *The Nation*, 28 May 2014
Winfrey, Oprah, 'Oprah Talks to Maya Angelou', *O, The Oprah Magazine*, May 2013 and December 2000
Younge, Gary, 'Maya Angelou: "I'm fine as wine in the summertime"', *Guardian*, 14 November 2009

28. Ernest Hemingway

'Chronicling Ernest Hemingway's Relationship With The Soviets', *Weekend Edition*, National Public Radio, 18 March 2017
'Ernest Hemingway', FBI Records: The Vault, https://vault.fbi.gov/ernest-miller-hemingway/ernest-hemingway-part-01-of-01/
'Was Ernest Hemingway a player in international espionage?', CBS News, 11 March 2017
Arnone, Terra, 'Ernest Hemingway's Crook Factory and other things we learned from Writer, Sailor, Soldier, Spy', *National Post*, 24 March 2017
Beaumont, Peter, 'Fresh claim over role the FBI played in suicide of Ernest Hemingway', *Observer*, 3 July 2011
Bradford, Richard, *The Man Who Wasn't There: A Life of Ernest Hemingway* (Bloomsbury, 2020)
Cowles, Gregory, 'The Great American Novelist Who Spied for the Soviets', *The New York Times*, 24 March 2017
Drye, Willie, 'The True Story of the Most Intense Hurricane You've Never Heard Of', *National Geographic*, 8 September 2017
Dugdale, John, 'Hemingway revealed as failed KGB spy', *Guardian*, 9 July 2009

Gilroy, Harry, 'Widow Believes Hemingway Committed Suicide; She Tells of His Depression and His "Breakdown"', Assails Hotchner Book', *The New York Times*, 23 August 1966, via TimesMachine

Hemingway, Ernest, 'Who Murdered the Vets?', *New Masses*, 17 September 1935, via www.marxists.org/history/usa/pubs/new-masses/1935/v16n12-sep-17-1935-NM.pdf

Hotchner, A. E., 'Hemingway, Hounded by the Feds', *The New York Times*, 1 July 2011

Klehr, Harvey, 'Hemingway Was a Spy', *The Wall Street Journal*, 13 March 2017

Lewis White, Ray, 'Hemingway's Private Explanation of "The Torrents of Spring"', *Modern Fiction Studies*, vol. 13, no. 2, Summer 1967

Masciotra, David, 'Ken Burns' vicious Hemingway smear: PBS series totally ignores writer's lifelong leftist politics', *Salon*, 11 April 2021

McDowell, Edwin, 'Publishing: Pulitzer Controversies', *The New York Times*, 11 May 1984, via TimesMachine

Mellow, James R., *Hemingway: A Life Without Consequences* (Da Capo Press, revised ed. 1993)

Meyers, Jeffrey, 'A Good Place to Work: Ernest Hemingway's Cuba', *Commonweal*, 7 February 2015

Mitgang, Herbert, 'Publishing F.B.I. File on Hemingway', *The New York Times*, 11 March 1983

Reynolds, Michael, *Hemingway: The 1930s through the Final Years*, (W. W. Norton & Company, 2012)

Reynolds, Nicholas, 'How Russia Recruited Ernest Hemingway', *The Daily Beast*, 18 March 2017

Timberg, Scott, 'Hemingway drank too much: Our strange, macho romance with Papa's alcoholism', *Salon*, 21 July 2016

29. Yukio Mishima

Review of Hiraoka, Azusa, *My Son: Yukio Mishima* (1972), https://nihongobookreview.wordpress.com/

Abelsen, Peter, 'Irony and Purity: Mishima', *Modern Asian Studies*, vol. 30, no. 3, July 1996

Cather, Kirsten, 'Japan's most famous writer committed suicide after a failed coup attempt', *The Conversation*, 11 January 2021

Darling, Laura, 'Yukio Mishima', *Making Queer History*, 1 October 2017

Flanagan, Damian, 'Yukio Mishima: Saints and seppuku', *The Japan Times*, 6 January 2018

Flanagan, Damian, 'Yukio Mishima's enduring, unexpected influence', *The Japan Times*, 21 November 2015

Graham, Thomas, 'Yukio Mishima: The strange tale of Japan's infamous novelist', *BBC Culture*, 25 November 2020

Ishiguro, Hide, 'Writer, Rightist or Freak?', *The New York Review*, 11 December 1975

Makoto, Furukawa, and Lockyer, Angus, 'The Changing Nature of Sexuality: The Three Codes Framing Homosexuality in Modern Japan', *U.S.-Japan Women's Journal*, English Supplement no. 7, 1994

McAdams, Dan P., 'Fantasy and Reality in the Death of Yukio Mishima', *Biography*, vol. 8, no. 4, Fall 1985

Mishima, Yukio, trans. Bett, Sam, 'Yukio Mishima on the Beautiful Death of James Dean', *Literary Hub*, 29 April 2019

Raeside, James, 'The Spirit Is Willing but the Flesh Is Strong: Mishima Yukio's "Kinjiki" and Oscar Wilde', *Comparative Literature Studies*, vol. 36, no. 1, 1999

Wagenaar, Dic, and Iwamoto, Yoshio, 'Yukio Mishima: Dialectics of Mind and Body', *Contemporary Literature*, vol. 16, no. 1, Winter 1975

30. NASA

'Nasa's disastrous year', BBC News, 22 March 2000

'Overview of the DART Mishap Investigation Results', NASA, 2005, www.nasa.gov/pdf/148072main_DART_mishap_overview.pdf

'The imperial units error that downed a multi-million-dollar Mars probe', *New Scientist*, 2 October 2019

'Where do astronauts go when they need "to go?"', *American Physiological Society*, 10 July 2013

Barry, Patrick L., 'Metric Moon', NASA, 8 January 2007, https://science.nasa.gov/science-news/science-at-nasa/2007/08jan_metricmoon

Beam, Christopher, 'Do Astronauts Have Sex?', *Slate*, 7 February 2007

Chang, Jon M., 'The History of Urinating in Space', ABC News, 11 July 2013

Evans, Ben, '"Man, I Gotta Pee": 55 Years Since Freedom 7 Began America's Adventure in Space (Part 1)', AmericaSpace, 30 April 2016

Groshong, Kimm, 'Spacecraft collision due to catalogue of errors', *New Scientist*, 16 May 2006

Grossman, Lisa, 'Nov. 10, 1999: Metric Math Mistake Muffed Mars Meteorology Mission', *Wired*, 10 November 2010

Hollins, Hunter, 'Forgotten hardware: how to urinate in a spacesuit', *Advances in Physiology Education*, vol. 37, no. 2, June 2013

Joshi, Shamani, 'Can Astronauts Masturbate in Space? An Investigation', *Vice*, 1 June 2020

Lloyd, Robin, 'Metric mishap caused loss of NASA orbiter', CNN, 30 September 1999

Marks, Paul, 'NASA criticised for sticking to imperial units', *New Scientist*, 22 June 2009

Rabie, Passant, 'NASA: 100 tampons and 2 other times space travel failed female astronauts', *Inverse*, 18 October 2019

Ride, Sally K., interviewed by Rebecca Wright, NASA Johnson Space Center Oral History Project, 22 October 2002, https://historycollection.jsc.nasa.gov/JSCHistoryPortal/history/oral_histories/RideSK/RideSK_10-22-02.htm

Schweickart, Russell, talking to Warshall, Peter, 'There Ain't No Graceful Way: Urination and Defecation in Zero-G', *The CQ*, Winter 1976–7, via https://space.nss.org/settlement/nasa/CoEvolutionBook/SPACE.HTML

Shira Teitel, Amy, 'Moon Diapers and Pee Condoms: the Evolution of Deep Space Evacuation', *Vice*, 17 September 2011

Taub, Amanda, 'NASA thought Sally Ride needed 100 tampons for 1 week "just to be safe." From what?', *Vox*, 26 May 2015

Trendacosta, Katharine, 'That Time Penis Bravado Caused NASA to Change its Condom Sizes', *Gizmodo*, 23 March 2014

Epilogue

1997 Annual Meeting, Warren Buffett Archive, https://buffett.cnbc.com/1997-berkshire-hathaway-annual-meeting/

Acknowledgements

As a kid, I could never understand why Oscar speeches went on so long. I vowed that if I were ever to receive an award or create some wonderful achievement that I was super proud of, I would do the opposite: just go balls to the wall, 'glad you liked it; I've nobody to thank but myself; peace out losers.'

Having now actually achieved something of note, however, I must begrudgingly admit that there are many, many people that it simply wouldn't have been possible without. Martyn: you've basically been a single parent for the last six months, and I owe you an enormous (**metaphorical** and **not legally binding**, because I know what you're like) debt for that. I'd say it'll never happen again, but we both know I've enjoyed becoming a real-life author too much for that to be the truth. Sorry.

On a practical level, I've simply got to mention Jom Felmtom, and Doug Young, both of whom were critical to the project's success and indeed existence at all. Doug: thank you for taking me on and supporting me and my absolute lack of business nous throughout everything.

To Lauras M and B, whom I can almost guarantee will not ever see this, I will be forever grateful. Without their expertise and encouragement, I may never have been able to start writing again after R was born; thank you both for reminding me that I existed as a person, outside of that dark fog of PPD, and that I could again.

Thank you to my mum, who instilled in me a love of reading, and to my grandad, who gave me a love of stories – they overlap, but they're not the same thing. The first stories I remember writing were on my grandad's PC – his old MS-DOS machine in his office that smelled like chess boards and tobacco – and I hope he'd be especially happy that I managed to get the story of Galois out to more people.

I already gave James a shout out, but it would be remiss of me to not include the rest of what my scrawled notes here appear to call 'the Shrumps': V, Dani, bex, Sorch, Rob, Boydy and Craig. Thank you for being my internet siblings, father and occasional drop-in meme machine all these years.

And last, but definitely not least, I am contractually obliged to thank my editor, Lindsay Davies.

Ha! Got ya. Lindsay is amazing: she was always there for me, endlessly supportive and understanding, and has, at this point, combed through hundreds of thousands of words of my bullshit so that you don't have to. I really couldn't have asked for a better editor, and I hope she doesn't mind my little joke here.

Thanks for reading, if you did. Sláinte motherfuckers!

Index

Aberdeen 128
Adams, John Couch 154–64; child prodigy 155; education 155–6; investigation of Uranus's wonky orbit 156–61; recognition for discovery 164; robbed 161–4
Airy, George Biddell 43n, 157–61, 162, 163, 164
Akamu, Nina 34–5
alchemy 74–5
Aldrin, Buzz 298
Alexander I, Tsar 121
altitude sickness 189–90
Amadeus (film) 84–5, 87
Amadeus (play) 92
American Crusade to End Lynching 249
American Magazine, The 224, 225, 227
Amsterdam 65–6
Analytical Engine, the 140–1
Analytical Society, the 81
Andreae, Johann Valentin 69
Angelou, Maya 263–73; accomplishments 263–4; birth name 265; early life 265–6; *I Know Why The Caged Bird Sings* 265; kidnapped 271–3; rape 266–7; San Francisco streetcar job 267–70
Aristotle 11, 12–13, 22, 63
Armstrong, Neil 298
asphyxia 191–2
astronomy 36–8, 42–8, 54–5, 58, 59, 154–64
Athenaeum, The 163
atmospheric science 193

Babbage, Charles 140–2, 160
Baillet, Adrien 66, 67
ballooning 183–94
batteries 95
Baudouin, Philippe 226
Bauer, Edgar 166–8, 170
Bauer, Josef 201

Baxter, Vivian 266–7, 267, 269, 270–3
Beagle, HMS 175–80, 181
beans 12–14
Becquerel, Henri 238, 240
Bell, Joseph 208–9
Berlin 113–15
Berlin, University of 166n, 170–1
Bernoulli numbers 141
Berthier, Alexandre 122, 123n, 124
Big Bang Theory 259
biographies 4
Blackwood's Magazine 135
blasphemy 38
Bologna, University of 39–40
bombardier beetles 177
Bonaparte, Napoleon 120–4, 148; brother 123; flees rabbits 124; military career 121; most famous quotation 120; rabbits defeat 122–4; stature 120
Bonn, University of 169–70
Borgia, Cesare 32
Bowie, David 230
Brahe, Beate 50
Brahe, Jørgen 50–1
Brahe, Otte 50, 53
Brahe, Tycho 49–61; *Astronomiae Instauratae Progymnasmata* 58; autopsy 60–1; birth 50; brass nose 49, 53; burials 60–1; death 59–60; devotes life to science 51–2; drunken pet moose 55–7; education 51; gene pool 50; and Kepler 58–60; kidnapped 50; levels of gold in body 60–1; loses nose 52–3; observatory-cum-palace 55; psychic dwarf 57–8; status 49; tour of Europe 54–5; wealth 50, 53
Breteuil, Louis Nicolas Le Tonnelier de 101n
British Association for the Advancement of Science 186
British Meteorological Society 185
Brun, Jacques de 103

Buffett, Warren 306
Butler, Nicholas Murray 276
Byers, Eben, death by Radithor 241
Byron, George Gordon Byron, 6th Baron 125–36; affair with Caroline Lamb 7, 132–3; affair with Countess Teresa Guiccioli 134; burial 136; at Cambridge 130; charisma 128; childhood 126–7; children 133, 134; club foot 127, 130, 131; death 136; *Don Juan* 130, 131, 134–5; father 126; first modern celebrity 132; fling with Lord Grey de Ruthyn 129; and the Greek revolution 135–6; at Harrow 129–30; inherits title 128–9; mad, bad, and dangerous to know 132; marriage 133–4, 137–8; mother 126–7, 128, 129; pet bear 130; status 136

Café Gradot 104–5
Calculus Controversy, The 80–1
California 267–72
Cambridge University 130, 155–6, 157–8, 175
Cardano, Gerolamo 40
Cartesian coordinates 63
Castlereagh, Lord 131
cat piano 117–18
Catherine of Siena, St 113n
Catholic Church 30
Chagas disease 181–2
Challis, James 160
charisma 10
Charles the Affable, King of France 30
Charles X, King of France 148
Chastellet-Lomont, Marquis Florent-Claude du 104, 107
Châtelet, Gabrielle Émilie Le Tonnelier de Breteuil, Marquise du 100–9; affair with Voltaire 106–8; background 100–1; biggest weakness 102; card sharping 104; children 104, 108; con artist 107; death 108; duel 103; father 101; invents derivatives trading 107; marriage 104; presentation at court 102–3; studies maths 104; translation of Newton 105–6, 108

Chiarugi, Vincenzo 116
chimpanzees, get drunk 197
China 18–25
Christina, Queen of Sweden 71
Clairmont, Claire 134
cocaine 196–205
cogito ergo sum 62, 63, 64, 71
Cold War 277–8
Coleridge, Samuel Taylor 197n
Collinson, Peter 94, 96
computer programmers 137
Confraternity of the Immaculate Conception 29–30
Confucius 18–25; *Analects* 22; appearance 19; birth 19; conservatism 22; convinced of own failure 19, 25; crappy jobs 21; death 24; doctrine of hard work 23; influence 18; social outcast 20; status 25; summarises his life 20; teaching career 21–5
Copernicus, Nicolaus 37–8, 43, 44, 45
coprographia 90
coprolalia 90
cosmic background radiation 259
cosmological models 259–60
Cottingley fairies, the 214–18
Covid-19 pandemic 75
Cox, Brian 36
Coxwell, Henry Tracey 186–94
Crosse, John 143n
Curie, Marie 4–5, 235–41; background 235–7; death 240; education 237; lack of recognition 240; marriage 237–8; medical studies 237; Nobel Prizes 238, 240; notebooks radioactive 240; work with radioactive materials 238–40
Curie, Pierre 237–8, 238–9, 240
Cutchogue, Long Island 2, 243, 244–5

Dampier, William 179
Dante 39n
dark ages, the 254
Darwin, Charles 175–82; culinary adventure 177–80; death 181–2; experiments on worms 181; *On the Origin of Species* 180; relationship with

FitzRoy 175–7; voyage on the *Beagle* 175–80, 181
Darwin's rhea 178
Davy, Humphry 220
de Vere Stacpoole, Henry 216
demonic possession 113
Demuth, Helene 172
Demuth, Henry Frederick 171–2
derivatives trading 107
Descartes, René 62–71; birth 62; breakthrough 64; cogito ergo sum 62, 63, 64, 71; *Discourse on the Method* 65, 67; education 64–5; and the existence of God 70–1; invention of graphs 63; *Mathematical Thesaurus of Polybius Cosmopolitanus* 68; night in the oven 66–8; Revelation 66–8; and the Rosicrucians 68–70; status 62
Devonshire, Duchess of 132
Dickens, Charles 198
Difference Engine, the 140–2
Dinwiddie, William Walter 227
Doyle, Arthur Conan 206–18; belief in fairies 207, 214–18; career paths tried 211; childhood 207; death 217; medical practice 210–11; medical studies 208–9; and Sherlock Holmes 206–7, 208–9, 211; and spiritualism 211–14
drapetomania 111n
drug use 196–205
Du Bois, W. E. B. 247
duels 52–3, 103, 145, 169n
  mathematical 40, 41n
Dumas, Alexandre 149
Durham, University of 2–3

Eckstein, Emma (Irma) 202–4
École Polytechnique 147, 148, 152
Edinburgh, University of 208–9
Edison, Thomas Alva 219–28; 'Diary and Sundry Observations' 226; *Electrocuting an Elephant* (film) 223; electronic vote-recorder 221; failures 224; genius 220; hype 222; iconic creations 219–20; lightbulbs 220–2; phonograph 221, 223; Spirit Phone 224–8

Edison Project, the 226n
Edison Test, the 222
Edleston, John 130
education 23
Einstein, Albert 242–53, 306; arrival in America 247, 248; banned from working on A-bomb 250–1; dumb as a box of rocks 1; flees Germany 248; and Gödel 258–61; Lincoln University address 249; love life 250–3; never learned to swim 1–2, 244; popularity 246–8; public nuisance reputation 2; sailing skills, lack of 242–5; security risk 252–3; status 242; theory of general relativity 247
Einstein-Szilard letter 249–50
electricity 93–9
electroconvulsive therapy 110, 110–11n
Engels, Friedrich 171–2
Enlightenment, the 112
Erlenmeyer, Friedrich Albrecht 200
eugenics 232

fairy photographs 214–18
Farmer, Moses G. 220
FBI 278–9, 282
female astronauts 298–301
female geniuses 5
Ferro, Scipione del 41n
Feynman, Richard 248
Fields Medal 16
Fior, Antonio 41n
First World War 225
Fisher, Carrie 110–11n
FitzRoy, Robert 174–7, 180
Fleischl-Marxow, Ernst von 199–201
Fliess, Wilhelm 203, 204
Florence 31–2, 32–3, 38, 39, 47
Flying Boy, the 95n
Flying University, Warsaw 237
France 100–1, 146–53, 184
Franklin, Benjamin 93–9; electrical experiments 94, 95, 96–8; electrocutes self 97; pranks 93–4, 96; status 93; thunderstorm and key experiment 97–9
Freud, Anna 205

Freud, Martha (nee Bernays) 196–7, 200, 205
Freud, Sigmund 119, 195–205; children 197; cocaine habit 196–205; courts Martha 196–7; death 205; and death of Fleishl 199–201; Eckstein episode 202–4; fame 205; flees Nazis 204–5; *The Interpretation of Dreams* 202; Irma's Injection dream 202; jaw cancer 204, 205; legacy 198, 205; number paranoia 201–2; work on cocaine 199
Freyberger, Louise 172
Fuckboy, The 125

Galileo Galilei 38–48; birth 38; *Dialogue Concerning the Two Chief World Systems* 46–7; *Dialogues Concerning Two New Sciences* 47–8; education 38–9; heliocentrism 43–8; house arrest 47; *La Balancitta* 39; lecture on Dante 39–40, 42; *Sidereus Nuncius* 44
Galle, Johann Gottfried 162
Galois, Évariste 144–53; birth 146; death 145, 152–3; education 146–7; own worst enemy 147; predicts death 150n; *Préface* 150–2; radicalism 148, 149; status 144–5; trial 149–50
Galt, John 126, 127
Gellhorn, Martha 275, 277
Geneva Conventions 281
geniuses; default 5–6; female 5; internet search 5; lucky bastards 4; shortcomings and accomplishments. 4–5
Germain, Sophie 149
Germany 68–70, 113–14
giant tortoises, extreme tastiness 178–80
Giraud, Nicolas 130
Glaisher, James 183–94; altitude reached 193; altitude sickness 189–90; asphyxia 191–2; descent 192–3; disastrous flight 184, 190–4; first balloon flight 186–90; later flights 194; marriage 185; status 184–5; *Travels in the Air* 185–6
Glenn, John 297
God 70–1

Gödel, Kurt 254–62; accidentally discovers time travel 259–60; background 256–7; cosmological model 259–60; death 261–2; demons 257; and Einstein 258–61; favourite film 256; flees Germany 258; innocence 261; paranoia 257, 261–2; and US Constitution 260–1; revelation 256; status 255–6
Gödel's Loophole 261
Goethe, Johann Wolfgang von 115
Golos, Jacob 275–6, 277
Good Will Hunting (film) 144, 149
Gray, Stephen 95n
Greek revolution 135–6
Greenwich Observatory 158, 160
Grey de Ruthyn, Lord 129
Griffiths, Frances 214–16, 217
Grissom, Gus 294, 296
group theory 146
Guiccioli, Countess Teresa 134

Habsburgs, the, limited gene pool 50
Hall, Peter 92
Halle University 119
Handel, George Frideric 167n
Harrow 129–30
Hawking, Stephen 157
Hazlitt, William 136
heliocentrism 37–8, 42–8
Hell, geography of 39–40
Hemingway, Ernest 274, 274–83; contribution to the war effort 278–82; depression and paranoia 274; FBI investigation 274, 282; intelligence operations in Cuba 278–80; life a fucking mess 275; Nobel Prize for Literature 282; suicide 274, 282; *The Torrents of Spring* 277n; *For Whom The Bell Tolls* 276; work for the NKVD 275–7
Henry VIII, King 125
heroin 198
Herschel, Sir John 163–4
Herschel, William 156n
Herz, Marcus 114
Hewlett, Maurice 216

Hilbert, David 255
Hippasus of Metapontum, Pythagoreans murder of 14–17
Hodgson, Francis 130
Holland 123n
Hoover, J. Edgar 278
Houdini, Harry 211–12, 213, 225
Hussey, the Reverend Thomas John 158–60
Hven 54–5
hydrogen 187
hysteria caused by too much masturbation, nose surgery as cure of 203

imaginary numbers 40, 40–1n
incompleteness 255
innovators 10
Inquisition, the 43, 45, 46–8
International Psychoanalytical Association 204
International Zionist Organisation 247
irrational numbers 14–17
Islamic world 17
Italy 134

Jena, University of 171
Jeppe, psychic dwarf 57–8
Johnson, Bailey, Jr, 265, 267
Jones, Ernest 204
Jørgensdatter, Kirsten Barbara 53

Kant, Immanuel 114, 115
Katzenklavier, the 117–18
Keats, John 131
Kepler, Johannes 44, 58–60, 79
Kepler, Katharina 58n
Kierkegaard, Soren 71
King, Martin Luther, Jr 264
Kinnersley, Ebenezer 220
Konenkova, Margarita 252–3

Lamb, Caroline 132–3
Lange, Joseph 91–2
Lao Tzu 24
laudanum 197

Laurent, François, Marquis of Arlanders 184
Le Verrier, Urbain Jean Joseph 161–4
Leddy, Raymond 278–9, 280
leguminophobia 12–13
Leibniz, Gottfried 76–7, 80–1
Leo X, Pope 33–4
Leonardo da Vinci 2, 4, 26–35; anatomical drawings 33; competition with Michelangelo 32–3; death 34; first commission 27–8; horse sculpture 28–9, 30, 33, 34–5; inventions 28; *The Last Supper* 31; *Mona Lisa* 32, 34; move to France 34; notebooks 33, 34; self-portrait 28n; status 26; *The Virgin of the Rocks* 29–30
Leonardo of Pisa 40
lessons 305–6
Leyden jars 94–5, 98
Liebknecht, Wilhelm 166–8, 166n
life force 115–16
light 36
lightbulb, invention of 220–2
Lincoln, Abraham 197
Lincoln, Mary Todd 197
Lincoln University 249
Liouville, Joseph 146
London, Marx's pub crawl 165–9
London Guardian 163–4
Long, Edward Noel 130
Louis Philippe, Emperor 148
Louis XIV, King of France 101n
Lovelace, Ada 137–43; and Babbage 140–2; birth 133, 137; Byronic tendencies 139; childhood 138; death 143; gambling debt 139, 142–3; legacy 143; mysterious illnesses 139; Note G 141; pawns family jewels 142–3; scientific curiosity 138–9
luck 7, 306
Luther, Martin 36

McBarron, James 297
Machiavelli, Niccolò 32
Malcom X 264
Mandela, Nelson 264
Mandeville, Bernard 102

Manning, Robert 282, 283
Marič, Mileva 250–1
Marie Antoinette 86
Marriott, William S. 212–13
Mars Climate Orbiter mission fuck-up 301–3
Marx, Eleanor 171–2
Marx, Heinrich 169n, 170
Marx, Karl 165–73; children 171–2; duels 170; and Engels 171–2; father 169n; London pub crawl 165–9; reputation 172–3; shitty friend and husband 172; as a student 168–71
Masked Medium, the 213–4
mathematical duels 40, 41n
Menlo Park laboratory 219–20
mental illness 110–19, 132–3
Michelangelo 32–3
Mikhailov, Pavel 252–3
Milan 28–31, 33
Milbanke, Annabella 133–4, 138, 139, 142, 143
Mishima, Yukio 284–92; acting career 289; body-building regimen 289; *Confessions of a Mask* 287–8; double life 287; education 286; first novel 287n; first short story published 286; *Forbidden Colours* 288–9; founds the Tatenokai 290; insurrection attempt 290–2; marriage 286–7; relationship with his mother 285; reverence for samurai lore 285; ritual suicide 291; *Temple of the Golden Pavilion* 289; upbringing 284–6
Moleyns, Frederick de 220
Montaigne, Michel de 102
moral management 117
morality 25
Morgenstern, Oskar 257, 260
Morita, Masakatsu 290, 291
morphine 198, 199–200
Mozart, Leopold 85–7
Mozart, Nannerl 85–7, 91
Mozart, Wolfgang Amadeus 84–93; arse obsession 89–90, 91, 92; canon in B-flat major 89; character 84–5, 87, 91–2; child prodigy 85–7; daddy issues 86–7; death 92; death of mother 86–7;
Difficile lectu, K. 559 91; motor tics 90; musical legacy 84, 85; proposes to Marie Antoinette 86; sense of humour 89–90, 91; Tourette's syndrome 90; wants to feed first born water 87–9
music, as therapeutic tool 118

NAACP 249
NASA 293–304; DART mission 303; Disposable Absorption Containment Trunks 298–9; female astronauts 298, 299–301; Group 8 class astronauts 299–300; Mars Climate Orbiter mission fuck-up 301–3; urine collection systems 293–9; use of two systems of measurement 301–4
Nazis; annexation of Austria 204-5, 258; antisemitism 258; bounty on Einstein 248; come to power 204; fucking idiots 246; potential to build nuclear bomb 246, 249; seize Einstein's yacht 243
Neptune, discovery of 154–64
neuroscience 2
New York Times, The 252–3
New Zealand 174–5
Newton, Isaac 72–83; alchemy 74–5; apple story 72; birth 73; the Calculus Controversy 80–1; doesn't know not to stare directly at the sun 3–4, 77–8; epitaph 73; laws of motion 78; as Master of the Royal Mint 81–3; mistake 79–80; *Philosophiæ Naturalis Principia Mathematica* 78–80, 105–6, 108; pokes needle in eye 76–7; as psychopath child 73; search for the Philosopher's Stone 74–5; status 72–3, 81, 83; year of wonders 75–81
Newtonmas 73
Nietzsche, Friedrich 71
ninja turtles 27
NKVD 275-7
Nobel Prize 238, 240; none for mathematics 16n
nose surgery, as cure for hysteria caused by too much masturbation 203

nuclear weapons 245–6, 249–50

Oasis 83
O'Hara, Dee 296
Olbers' paradox 36
Oppenheimer, Robert 252–3
Osiander 37–8

Packard, Elizabeth 111n
Padua, University of 42
pandemic, 1918 225
Paris 171, 184, 237–8, 281
Parks, Rosa 268
Parsberg, Manderup 52
Pennsylvania Gazette 99
person from Porlock, the 197n
Philosopher's Stone 74–5
philosophy 62–71
phobias 12–13
Pilâtre de Rozier, Jean François 184
Pinel, Philippe 116–17
Pisa, University of 42
Plague, the 75
Planck, Max 242
Plato 22, 63
Pleasant, Mary Ellen 6
Pluto 154, 164
Poisson, Siméon Denis 151
Poland 235–7
polonium 238–9
Porter, David 179
Prague 58
Prestige, The (film) 230, 231n
Priestley, Joseph 98
psychiatry 116–19, 195
Pyramids, the 75
Pythagoras 9–14; charisma 10; cult of 10–11; cultists murder of Hippasus of Metapontum 14–17; death caused by beans 12–14, 17, 145; declared dead 12; leguminophobia 12–13; mythology 10; theorem 9, 10

rabbits, defeat Napoleon 122–4
racism 248–9, 263–73
radioactivity 238–41

Radithor energy drink 241
radium 238–9, 240–1
Ramanujan, Srinivasa 6
Raspail, François-Vincent 150n
Reil, Johann Christian 113–19, 195; invention of psychiatry 116–19; *Katzenklavier* 117–18; prescription of sex and drugs 118
Rethke, Donald 298
Rheticus, Georg Joachim 37, 44
Rhode Island 243–4
Ride, Sally 300–1
right to fuck up 7
Robeson, Paul 249
Rosenkreuz, Christian 69
Rosicrucians 68–70
Rostock, University of 52–3
Rothman, Robert 243
Royal Mint 83
Royal Society 76, 162
Russell, Bertrand 256
Russia 121

safety protocols, ignoring 235–41
Saint Louis 265
Salieri, Antonio 84, 87
Salk, Jonas 248
San Donato a Scopeto 28
San Francisco 267–72
Santa Maria delle Grazie, convent of 31
Scarface (film) 195–6, 199
scepticism 62
Schiller, Friedrich 115
Schlick, Moritz 257–8
Scholasticism 63
Schur, Max 202, 205
Schweickart, Rusty 297–8
scientific method 62
scientific observation, precedence over Biblical lore 45
scientific revolution, the 254–5
Second World War, Hemingway's contribution 278–82
sex in space 293–4
Sforza, Ludovico, Duke of Milan 28, 30
Shelley, Mary 134

Shelley, Percy Bysshe 134
Shepard, Alan 294–6
shortcomings, accomplishments and 4–5
Shriver, Jean 249
slaves and slavery 111n, 174, 176–7
Snow White and the Seven Dwarves (film) 256
Social Democratic Party of Germany 166n
sodomy 130n
solar system 154–64
South Sea Bubble 83
Southey, Robert 131
Southwell, Robert 77
Spears, Britney 85
spirit photographs 212–13
spiritualism 211, 225–6
Stamps, Arkansas 265, 267
static electricity 94
Stevenson, Robert Louis 209
Stevin, Simon 66
Strand, The 215, 218
Sudoplatov, Pavel 252–3
Swan, Joseph 220
Sweden 71
Swezey, Kenneth M. 231
Switzerland 134
syphilis 118
Szilard, Leo 246, 249

telescopes 42, 44
Tesla, Nikola 229–34; ahead of his time 229–30; birthday celebrations 231–2; death 234; death ray 232–3, 234; Experimental Station, Colorado Springs 231; love of pigeon 233–4; played by David Bowie 230
Thatcher, Margaret 92
Thiébault, Baron Paul Charles François Adrien Henri Dieudonné 122, 123, 124
Thompson, J. J. 247
Thompson, Louise 243
Tierra del Fuego 176
Tilsit, Treaties of 121
time travel 259–60
The Times 188

Tourette's syndrome 90
tropopause, the 193
Trump, Donald J., stupidity 3–4
Tuke, William 116

United States of America; Great Labor Day Hurricane, 1935 276; racism 248–9, 263–73; Red Scare 277–8
Uraniborg 55
uranium 238, 239–40
Uranus, wonky orbit 154–5, 156–62
Urban VIII, Pope 45–6, 47
US Constitution, inner contradictions 260–1

Vasari, Giorgio 26, 27n
Vellutello, Alessandro 42
Venice 31
Venus 43
Verrocchio, Andrea del 27
Victorian era, weird time 183–4
Volosin, Jeff 304
Volta, Louis 95
Voltaire 106–8

Wall Street Journal 241
War of the Currents 223, 224
War of the Fifth Coalition 122
War of the Fourth Coalition 121
Warsaw 235–7; Flying University 237
Watters, Leon 244
Wednesday Psychological Society 204
weirdo geniuses, types 4–5
Weizmann, Chaim 247
Westphalen, Jenny von 169n, 171, 171–2, 172n
Wilhelm IV, Landgrave of Hesse-Kassel 56
witches 58n, 113
women; freedom 100–9; mental illness 111
Wordsworth, William 131
Wright, Elsie 214–16, 217

X-rays 238